The CLOCK REPAIRER'S HANDBOOK

The CLOCK REPAIRER'S HANDBOOK

Laurie Penman

SKYHORSE PUBLISHING

Skyhorse Publishing books may be purchased in bulk at special discounts for sales promotion, corporate gifts, fund-raising, or educational purposes. Special editions can also be created to specifications. For details, contact the Special Sales Department, Skyhorse Publishing, 307 West 36th Street, 11th Floor, New York, NY 10018 or info@skyhorsepublishing.com.

Skyhorse® and Skyhorse Publishing ® are registered trademarks of Skyhorse Publishing, Inc. ®, a Delaware corporation.

Visit our website at www.skyhorsepublishing.com.

10 9 8 7 6 5 4

Library of Congress Cataloging-in-Publication Data
Penman, Laurie.
The clock repairer's handbook / Laurie Penman.
p. cm.
Originally published: Newton Abbot [Devon] : David & Charles ; New York :
Arco Pub., c1985.
Includes bibliographical references and index.
ISBN 978-1-60239-961-7 (alk. paper)
1. Clocks and watches--Repairing. I. Title.
TS547.P4 2010
681.1'130288--dc22
2010010216

Printed in the United States of America

Contents

Introduction

This book has been written with the intention that it should prove useful to both the absolute beginner and the more experienced clock repairer, without baffling the former or annoying the latter. I hope that the intention is realised.

Please read Chapter 1 first. If, like me, you are in the habit of skipping through a book looking for particularly interesting parts, restrain yourself. This first chapter gives an overall coverage of repairs and it tries to persuade you to tackle them systematically. Thereafter the book deals with the subject area by area and you are invited to jump directly to whatever covers the problem of the moment, but a disciplined approach to diagnosing faults is well worth cultivating. Whether you pursue the craft for gain or for pleasure, I believe that you will profit from avoiding the 'dive straight in' method, tempting though it is.

For those of you who are new to clockmaking, let me assure you that it is *not* necessary to invest in a workshop and an expensive collection of tools, good work can be done at the kitchen table with simple, good household tools — just take care and apply a lot of consideration to what is probably the oldest and most loved domestic machine in our homes.

Enjoy yourself!

Laurie Penman
Totnes, Devon

1
Finding Clock Faults

This first chapter is intended as a quick reference aid for solving clock repairing problems. Most of the solutions are not set out in great detail – the other chapters do that. Here I simply try to help you to locate the problem and to give a little help in making what may be perfectly obvious corrections. On occasion you will find that there are two solutions proposed, one in this chapter and another in the specialised chapter. There is frequently more than one good solution to a problem and this is a convenient method of showing them.

FAULTS COMMON TO MOST CLOCKS

Diagnosis

The first necessity is diagnosis. Often the locality of the fault is clear: a hand is catching on the dial, or a gear is badly damaged, for instance. But this is not always so and it is worthwhile adopting a simple pattern of testing that will assist in finding faults. Of course the clock may be so dirty that it does not have a hope of going until the dirt and old oil has been removed. In this event it is useful to take the movement out of its case and carry out a crude (but safe) 'dunking' in paraffin (kerosene) that contains about 5 per cent of good lubricating oil. After draining it should be possible to proceed with the following tests. Do remember that this is *not* a proper cleaning; it is simply a means of loosening-up the clock for testing.

Consider the clock as a series of systems connected together and test each system in succession. The first thing to test is the power. Is the clock wound up? Is the mains or battery supply making a proper connection? Then test as follows:

1. Open any part of the case that can, conceivably, come into contact with any part of the movement (including weights and pendulum).
2. Check to see if the clock will run now. This check follows each test.
3. Remove the hands or disconnect the display.
4. Remove the dial.
5. Remove the pendulum. Many clocks will need the crutch adjusted to put it 'in beat'. Recoil escapements will run without anything further being done, but dead-beat escapements need a little weight added to the crutch so that it will unlock the escape wheel. Blu-tack or a similar stick-on product performs this service very well.
6. Remove the hour and minute pipes.

If the clock runs after carrying out any one of these tests, the fault lies in the part that has just been removed. In a movement that has been working for years, you will most probably have to carry on and strip and clean the complete movement in a proper fashion (see Chapter 2). The tests should ensure that you do not waste time in correcting parts that look suspicious but in fact are working satisfactorily, or in taking a movement apart when the fault is caused by some exterior factor. After repairing the movement there is sometimes a disappointing lack of response from the mechanism. Repeat this series of tests – do not assume that there is an undiscovered fault in the movement until you have proved it, for life is too short to strip clock movements unnecessarily. Besides it is not fair on the clock. It is always best *not* to strip a clock if there is no sign of dirt, stiff oil or obvious wear or corrosion. The wheels and pinions will have worn together over the years and if the wheel count of teeth is precisely divisible by the pinion count, each leaf will have its own set of wheel teeth that it complements. Changing the set of wheel teeth that each leaf meshes with, will leave a meshing of wheel and pinion that is not as good as existed before dismantling. (This is the main reason for filing or stoning out any 'pocketing' of the pinion leaves.) A movement does not need dismantling if it operates and:

It is clean, with no stiff oil or 'varnish'.
It has slight pocketing but no evidence of black or metallic particles.
It has sloppy holes that are round, with no evident wear of the pivots.
In those few movements that hold the escapement pallet arbor between the clock plates, if:
the escape pallet arbor does not lift as the wheel rotates.
In longcase or grandfather clocks, if they:
stop at the same day each week (eight-day clocks).
stop at the same hour each day (thirty-hour clocks).
These last two faults typify those that disappear when the case door is left open, and are a result of sympathetic vibration.

OPENING THE CASE

This first test should eliminate stoppages due to the hands fouling the glass and, in weight-driven pendulum clocks, any impedence to the steady fall of the weights. It will also allow you to see if the pendulum and weights touch at some part of the clock's going, or if there is any sympathetic vibration. All these will be fairly obvious, but sympathetic vibration is unusual and, when it occurs, quite difficult to spot. At a time when the weight cords have unwound to approximately the same length as the pendulum, there is a tendency for the weight or weights to oscillate in harmony with the pendulum bob. This may be sufficient to rock the seat board or case if they are imperfectly supported, or cause interference between pendulum, weights or case. In any event the clock will show the habit of stopping, for instance, every fourth day, or always at three o'clock in the morning. The cure for sympathetic vibration is to improve the support of seat board and case – in extreme cases associated with very heavy clock weights fasten the case to the wall – and to ensure that there is plenty of space between the weights and the bob at the time when they are almost level. Clocks often have two suspension positions (Fig 1), and moving the pendulum from one to the other of these will change the position of the bob in relation to the weights and the case. The position of the cord anchorage on the seat board will also affect the position of the weights in relation to the door or pendulum bob as the cord unwinds and the weight moves over in the direction of this anchorage.

I have mentioned sympathetic vibration and the need for a proper support for seat board and case without pointing out that improper support is a frequent fault in longcase clocks. Do not stand these clocks directly on thick carpet, or floorboards that cross doorways. If the clock *must* stand over a thick carpet, make a support for it by standing a board underneath that rests on three screws eased through the weave of the carpet and into the floor. In similar fashion do not fit shelf clocks or wall clocks to walls with doors let into them if you can avoid it, or walls that have a piece of vibrating equipment such as refrigerator, freezer or central-heating boiler resting against them.

REMOVING THE HANDS

Obviously removing the hands will cure any interference between the hands, but there are a few points that are not immediately apparent.

Occasional fouling of one hand on the other can be the result of using the wrong washer beneath the hand-retaining pin. The washer should be dished so that when it is pressed down by the pin its outer edge bears on the hand rather than any other part of its surface resting on the minute pipe. If the washer bears on the pipe, the minute hand will most probably be unstable, even though it may feel firm (Fig 2). The hour hand and pipe should not rub against the back of the minute hand; this is often prevented by the design of the cock that supports the compound wheel (the minute wheel), or the post that performs the same task. The hour wheel is trapped either by the overhang of the cock or by a washer under the retaining pin on the post. If there is no room for a washer, the taper pin alone will do the job if it is sufficiently long and positioned to reach past the root of the teeth (Fig 3).

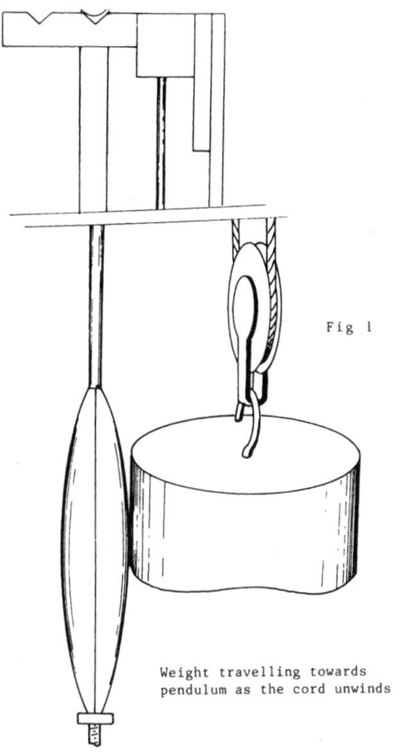

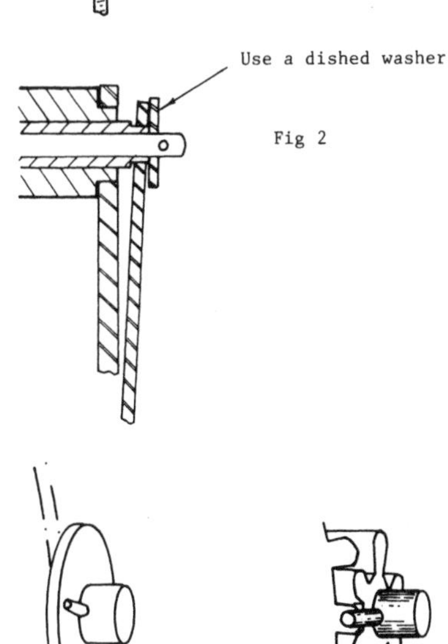

REMOVING THE DIAL

Old clocks with heavy dials place a great deal of strain on the support pillars; and if these allow the dial to droop, the hour pipe can be fouled quite enough to stop the clock. The single plate of a Victorian period dial is particularly liable to do this because the pillars were very often held in place by small screws and were not riveted into the plate. The fault can be corrected by filing out the hole that the pipes reach through, but this does nothing to remove the cause and, what is more, will give the impression that the dial has been 'married' to the movement and is not original. For a brass dial, either re-rivet using a polished planishing hammer and polishing and silvering afterwards if necessary, or use slightly larger screws and re-tap the pillars. Painted dials do not allow re-riveting, the best solution for them is to support the lower edge of the dial by attaching a piece of thin plywood to the uprights that carry the seat board, or any other form of support that is unobtrusive and does not mar the original work of the casemaker.

REMOVING THE PENDULUM

If the clock can be persuaded to go reliably when the pendulum is removed, the fault is either one already discussed or is caused by some idiosyncrasy of pendulum or suspension (which includes the cock). We must also consider the crutch. Faults and the necessary corrections to pendulums, suspensions and crutches are dealt with in detail in Chapter 9, but there are some simple points that can be looked at now.

The pendulum should not rattle anywhere – the connections between suspension and cock, suspension and rod, rod and crutch, rod and pendulum bob should be firm. In the case of the fit of the suspension in the suspension cock and the rod in the crutch there should be easy movement, but not looseness, between the two parts. Ensure that the path of the pendulum bob when viewed from above is at right angles to the crutch arbor and that it is not twisted on the rod so that it exposes more than its minimum cross-section to the direction of movement (Fig 4). A significant twist to the bob will give a sideways thrust to the pendulum and cause the pendulum to weave about instead of beating in a single plane.

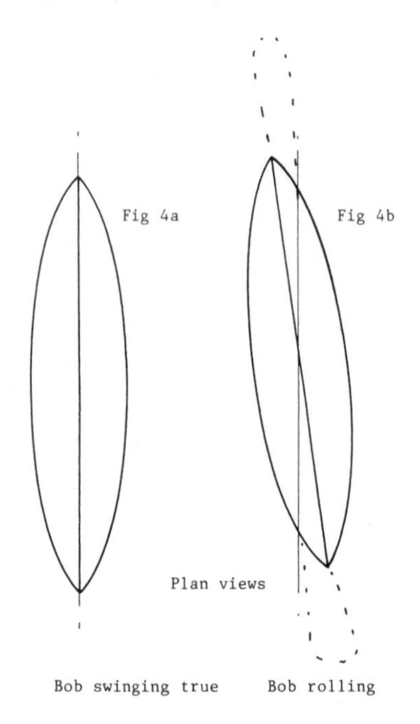

The suspension springs in longcase clocks are always much longer than is strictly necessary. If there is any sign of damage or corrosion and it is desirable to retain the original spring, it is quite acceptable to shorten it, making certain that the crutch will engage the rod, or extension to the rod, that is normally provided. No more – and probably a lot less – than 18mm (0.75in) of spring is needed for proper operation of the suspension.

REMOVING THE HOUR AND MINUTE PIPES

If the clock has been working for any length of time it is unlikely that parts of the pipes are rubbing to an extent that will cause failure; but dirt, old oil and airborne grit have a surprising affinity for the inside of pipes and often cause a clock to stop. Examine the mesh of the cannon, hour and minute wheels; damage as a result of people forcing the hand round often shows here and will stop the clock. Completely irregular faults, with no obvious cause, can often be traced to variable meshing as pipes move, or result from bent posts that present the minute pipe at an angle to the other wheels. I once discovered a broken tooth held in place by a thin flap of metal so that although it was capable of leaning over and jamming, it almost always flopped back again and presented an undamaged appearance when the dial was removed. It was only discovered when the faulty wheel was removed and handled. I have never known this fault to occur in the train wheels.

Escapement faults

When the previous tests have been made and a fault remains, one must turn to the rest of the movement. In most cases it is still not necessary to take the plates apart, because the escapement is fitted with a cock or is on a separate platform.

PLATFORM ESCAPEMENTS

Before removing the platform examine it closely, you will find a magnifying glass very useful. Give a sharp twist to the clock movement to see if the balance wheel will swing and the escape wheel rotate. If neither will move, there is either a broken pivot or the escapement is gummed up. Usually a broken pivot is obvious because if it is touched lightly with a thin piece of pivot wire, it will tip from side to side. Sound pivots allow the wheel to be lifted vertically and then drop again, without tilting. There are two types of common platform escapements – cylinder and lever – and these are illustrated in Chapter 6; only one, the lever, is available as a replacement. Repairs to a platform escapement are the province of the watch repairer rather than the clock repairer, but there are only a handful of craftsmen in Britain that are willing to accept repair work on these items. It follows then that if a cylinder escapement is broken you will almost always have to replace it with the lever type.

Proper cleaning of platform escapements requires dismantling the device, including the jewelled bearings; quite obviously this cannot be done when the platform is still in position on the rest of the movement. Before undoing the screws that hold the platform, make sure that the train wheels cannot turn by slipping a thin piece of pivot wire through the crossings of the wheel that engages the escape pinion.

Most good watch repairers will undertake to clean a platform escapement, or you can take advantage of a broken example and practise on it. Carbon tetrachloride is a good cleaner and there are several proprietary compounds available; lubrication with a good watch oil should be carried out before assembly. Since disassembling platform escapements makes use of watchmaking techniques this book contains only simple adjustments, but Further Reading includes titles that will be useful if you wish to carry out this type of repair.

ESCAPE WHEELS AND PALLETS

Set up the clock movement so that the escape wheel is being driven by its spring or weight. If all is well, the escape wheel will turn as the crutch is moved from side to side and the amount that each tooth moves before striking the impulse face (the incline plane that lifts the pallet) will be the same for the teeth entering the escapement as for those leaving it. This free movement is called the 'drop'; it should not be greater than about 10 per cent of the distance between two teeth. If you use a feeler gauge to measure this distance, bear in mind the fact that the gauge will be measuring from the tip of the tooth across the shortest distance to the escape pallet; for a recoil escapement this is *not* the drop, which should be measured as the length of a tangent from the tooth tip. If the angle of the impulse face is 45 degrees (see Chapter 5 for more information on the recoil escapement), the drop will be 1.4 times the gauged dimension. The drop for a dead-beat escapement can be measured directly with feelers.

If the pallets catch on one or more teeth, the space between the teeth is uneven (varying pitch), either because of inaccurate manufacture or because the tooth tips have become bent. Inaccurate manufacture can be ignored as a reason for the clock stopping if the escape wheel is original, or at least has clearly been working for some years. The clock is obviously accustomed to coping with the error. Bent teeth can be put right with a pair of flat-nosed pliers. Gently pinch the tip so that one jaw of the pliers rests half-way down the curved side and the other lies flat on the radial side (Fig 5). When the tip bends towards the radial side, which is the most common state, a slight squeeze of the pliers will bring the tip upright. If it leans towards the curved side, the pliers must be rocked over until the tip is vertical again.

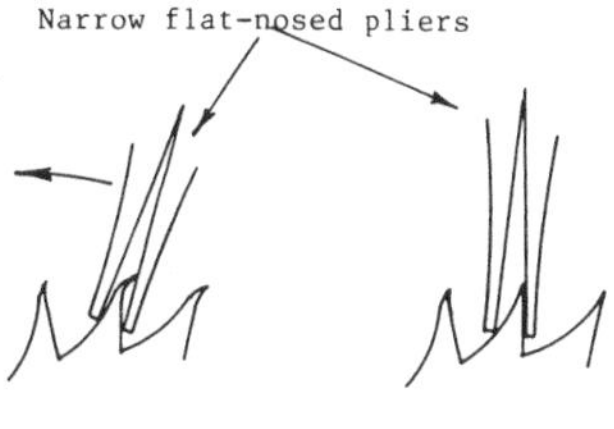

An escape wheel that passes some teeth through the escapement more easily than others has a great deal wrong. This cannot be treated without dismantling the movement completely in most clocks. There is nothing very complex about this, but this would seem to be the place to make three relevant points. Weight-driven clocks have the pendulum removed before the weights. Spring-driven clocks *must* have the spring let down and the ratchet or the click removed before attempting to dismantle the movement. If you are not familiar with the clock, make a sketch of the position of all parts as you take them off.

To correct an escape wheel that works unevenly, take the wheel and arbor out of the movement and turn it – or have it turned – until all teeth are the same height. The method of holding the wheel for turning is similar to that shown in Chapter 4 when machining the seating for a wheel. If the job is done properly, the shortest tooth will just have scratches on it from the turning tool and will show some of the old tip surface. The tooth tips will now be too thick; use a half-round file to remove metal from the curved side of the tooth until the tooth thickness has been reduced to leave a flat on the tip 0.1mm (0.004in) to 0.2mm (0.008in) wide. Do not touch the side that is radial or (in the case of the dead-beat escapement) nearly radial; this is probably still accurate so far as pitch is concerned (Fig 6).

Replace the wheel and arbor in the plates to test that it is rotating evenly beneath the pallets. It is not, however, sufficient to correct the wheel, because having altered this the pallets will no longer suit.

Narrow flat-nosed pliers

Fig 5

Straightening escape wheel teeth

Fig 6

File new curve

CORRECTING PALLETS

The pallets are driven by the escape wheel, and the impulse is transferred to the pendulum by the crutch. Since the latter is pivoted quite close to the point of flexure in most clock suspension springs, it will swing through approximately the same arc as the pendulum. For a longcase recoil escapement, this should be about 3 degrees on either side of the vertical. Keep contact between the escape wheel and the pallets and slowly rotate the wheel; mark the swing of the crutch with a soft lead pencil. The angle between the two extremes of the swing ought to be about 6 degrees – degree measurement is difficult to apply but 6 degrees is equivalent to a chord of 11.23mm (0.442in) on a crutch length of 100mm (3.937in) as shown in Fig 7. A chord measurement of between 9.5mm (0.375in) and 12.5mm (0.5in) is acceptable with this crutch length.

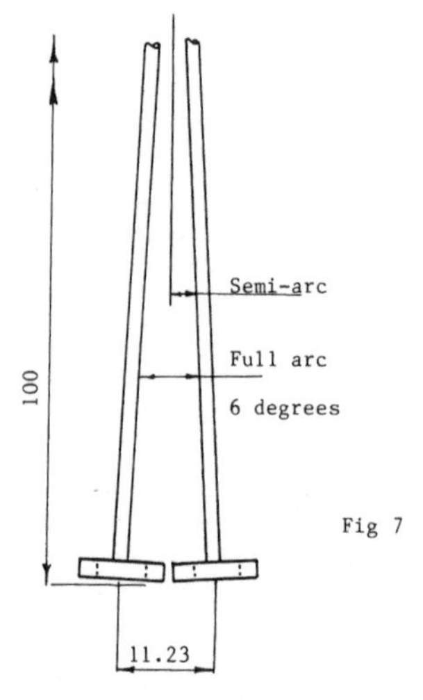

Fig 7

Proportions of crutch and arc

Any correction of the pallets is best carried out by drawing the pallets on card, using the wheel diameter and centre distance between the wheel and pallet arbors. If the wheel diameter or the centre distance has been altered from the original, use the new dimensions. The instructions for constructing recoil and dead-beat pallets are detailed in Chapters 5 and 6 respectively. When taking the centre distance remember that the arbors are not necessarily parallel and make the measurement by means of a vernier calliper held close against the escape wheel and pallets and making allowance for the diameters of the arbors.

After drawing the pallets, punch a hole on the escape-wheel centre so that the card can be slid over the pallet arbor and you can judge how far the pallets need to be bent and how much needs to be added to the impulse surfaces for them to match the drawing. If additions are to be made to the pallet faces, shape them before attachment so that little work needs to be done after soldering; harden them by heating to red heat and quenching in water in order to make the cleaning for soldering easier; tin the surface using solder, flux and a soldering iron and then tin the old surface of the pallets. Put the new piece in position on the pallet, tinned surface to tinned surface, then bring to soldering temperature by resting a hot soldering iron on the new piece or slowly heating with a gentle (quiet) gas flame. The soldering temperature will temper the new pieces to blue and make them sufficiently soft for a saw-sharpening file to shape them to suit the drawing. Polish after filing with emery and crocus paper.

SIMPLE RE-FACING OF PALLETS

A more common repair than the rather complex matter of turning the escape wheel and re-shaping the pallets, is straightforward compensation for the wear that takes place on recoil pallets. The evidence for this is a visible pit in the impulse face and the failure of the escapement to move the crutch through an acceptable angle (Fig 6). The wear can be corrected by removing metal from the impulse surface until a slip of spring steel can be soldered on and restore the original level of that surface.

Before you start, make sure that the remains of the original surfaces would give an even drop and the correct angle of crutch movement. If the drop is the same for both sides and the thickness of the spring steel that you have available is, for instance, 0.5mm (0.02in), it is only necessary to soften the pallets and file away 0.5mm (0.02in) from them. Do not worry about removing all evidence of the pits. If the original drop was uneven, now is the time to correct it by filing. Checking this matter of the original drop can be done quite simply by using feeler gauges

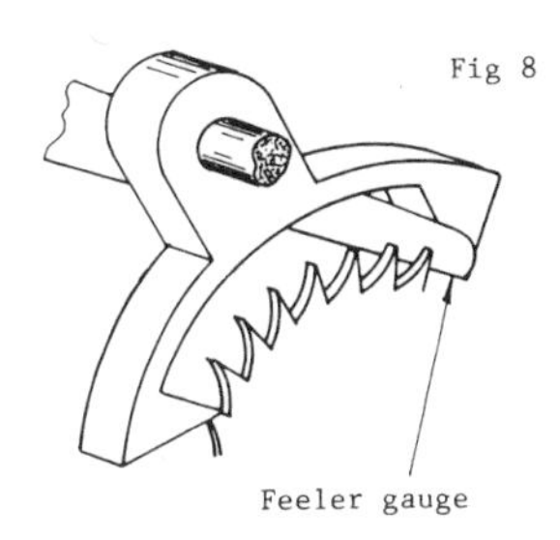

Measuring the drop

to bridge the pit and discover the drop on each pallet; do not forget that the actual drop will be 1.4 times this measurement (Fig 8). Tin and solder the spring-steel pieces as before.

Measurement of the amount of metal being removed is difficult. You can do it by putting the work back into the movement plates and using the feelers, or by measuring from impulse face to some other part of the anchor. A third method is to lay the anchor on a piece of scrap brass plate, locate it in place with a peg through the arbor hole and two other pegs registered on the outside, and record the starting position of the impulse faces by clamping two small strips of metal on the plate and resting against the faces.

CRUTCH COLLET

It is always a good idea to check the firmness of the attachment of the crutch to the pallet arbor. A slight failure of riveting or solder will allow movement of the crutch and prevent the full impulse being transferred to the pendulum.

Going train

Remove the escape pallets and try the freedom of the train. Do this by applying pressure to the teeth of the great wheel or going barrel until the train just turns, then very lightly touch the escape-wheel arbor with the tip of a finger so that there is the sort of back pressure on the train that the escapement would give. You will feel an unevenness to the rotation of the train, but it must not be such that would require much increase in drive (from the great wheel), to keep the escape wheel turning. If the train is alternatively hard and easy to keep going, the mesh or the pivots are at fault.

PIVOTS AND PIVOT HOLES

If the movement of the train is difficult, turn the train until the stiffest part of the rotation is found, stop turning and gently tip the plates to the horizontal. All the arbors of wheels and pinions that are meshing freely and have unbent pivots, should drop so that their shoulders rest on the movement plate. Turn the movement over so that the arbors fall in the opposite direction. Those arbors that failed to move on either of the tilts are suspect, mark them and the position of the teeth that are in mesh at the time and dismantle the movement. You can expect to find bent pivots, or gummed up or damaged teeth. This same tilting test that finds arbors that will not drop but do not show any significant variation in rotating, should reveal worn pivots and/or holes, or gummed up pivot holes.

Having checked the pivots for freedom, test the wear between pivot and hole. Inspect the holes for ovality and try to judge the amount of movement that the pivot can make from side to side and which will affect the ease of train rotation in one of two ways. The first will already have been seen in the test of train freedom, when you turned the wheel and noted unevenness of rotation. If the holes are oval the arbors will move away from each other and the meshing of wheel and pinion will alter. When the ovality is too much, the gears will mesh very badly and feel rough when turned; this can be so extreme that butting results and the gears will not turn at all (Fig 9).

The second effect is an increase in frictional losses, but not necessarily any great unevenness of rotation. The ovality may not be very obvious, but a great deal of sideways movement (shake) of the arbor may nevertheless be present as a result of the pivot wearing deeply and to a greater extent than the pivot hole. At the centre of the pivot there will be a waist and the original diameter on either side of this gives rise to two annular areas that rub on both sides of the plate (Fig 10). Both faults require the movement to be stripped and the pivots and holes refurbished (see Chapters 3 and 4).

The degree of sideways movement or ovality that can be accepted by the train without failure will depend on the size of the gear teeth, and can be evaluated by establishing

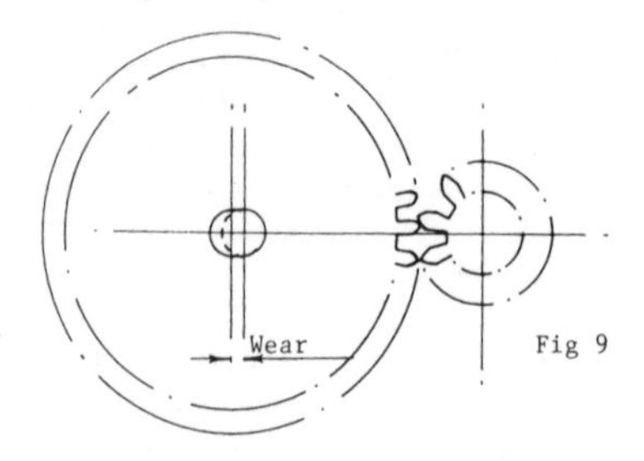

Fig 9

Bad meshing due to wear

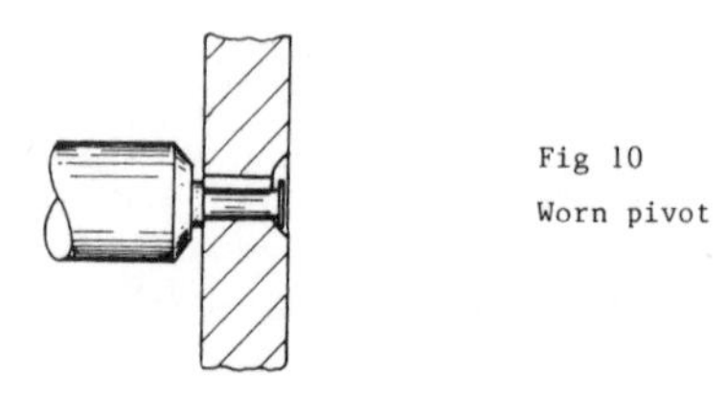

Fig 10
Worn pivot

the relationships between pitch circle, outside diameter and centre distances. Although gear teeth from a longcase or similar-aged clock will have been made to imperial measurements (or a contemporary system), the formulae used in calculating modern gears can be applied.

The pitch circles of meshing gears are equal to the number of teeth on each gear multiplied by the module.

$$PD = N \times Mod$$

Do not worry about the term 'module', that is simply defined by turning the formula around.

$$Mod = \frac{PD}{N}$$ where PD = pitch circle diameter
N = number of teeth

The outside diameters of gears can be found by adding an allowance to the number of teeth and multiplying that figure by the module. The allowance for wheels is different to that for pinions.

$OD=(N+f)\times Mod$ where OD = outside diameter
f = allowance
f = 2.71 for wheels
f = 1.71 for gears with less than 10 teeth
f = 1.61 for gears with more than 9 teeth

When a wheel and pinion are in mesh, the teeth overlap so that the pitch circles touch and the overlap is the sum of the addenda of both gears; the addendum being the distance from the pitch circle to the top of the tooth. Since this distance is half the difference between the OD and PD of the gear, the overlap can be found by multiplying the module by half the allowance 'f' for each gear. In British clocks this value is usually 2.15 to 2.5 times the module.

In practice the distance between centres can increase by up to 20 per cent of the overlap without butting taking place. The transmission of energy will not be as efficient as in a meshing pair whose pitch circles touch, but in most clocks the train will work. Twenty per cent can be taken as 0.5 x Mod, and if we consider a wheel and pinion of 0.8 Mod this allows a movement of the centres away from each other of 0.4mm (0.016in). This is extreme of course, but if we work within half this figure the movement can be expected to work without using weights heavier than usual or fitting a heavier spring, which is never acceptable. In a longcase clock this would be 3.5kg to 4.5kg (8lb to 10lb).

The above establishes a rule of thumb that allows 10 per cent of the module for the total movement of arbor centres. This will be reduced if there was any error in placing the arbor centres in the first place – the figure given is allowed on the theoretically correct positions of pivot holes.

The remedy for worn pivots is either to file them and then burnish as described in Chapter 4 or, if they would then be too slender, to remove them altogether and re-pivot which is also described in Chapter 4.

Worn pivot holes need bushing. However, remember that pivot holes in clock movements cannot wear round; if a hole is round it is unworn and it will only need to be bushed if the pivot is reduced in diameter. Chapter 3 describes bushing techniques.

WORN GEAR TEETH

Trains that are uneven in their rotation, with no evidence of worn pivots or pivot holes, are suffering from worn or distorted teeth. Both faults call for dismantling the movement and examination with a magnifying-glass.

Distorted teeth can only result from some mistreatment of the clock; had the fault been in the original manufacture, failure would have followed within a short time. A few distorted teeth on a wheel indicate that someone has continued heaving on the winding handle or key after the clock was fully wound (wounded is a reasonable description of an over-wound clock). In a spring-driven clock this can be caused by the click breaking or coming loose.

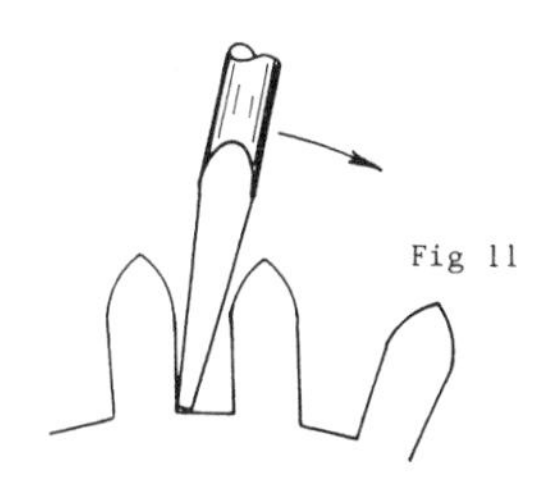

Straightening gear teeth

Teeth can be straightened by levering them upright with the blade of a screwdriver; rest the end of the blade against the bottom of an adjacent tooth and then press the bent one gently back into position (Fig 11). This does work sometimes, but there is always the risk of the tooth having been seriously weakened even though the fractures are microscopic and not seen through a magnifying-glass. Repair by breaking the tooth out and then sawing down the centre of the stub or scar with a saw blade that is slightly narrower than the tooth. File out with a flat file that has had one side ground smooth, and then fit and solder a new piece of 70/30 brass that has been hammered to the tooth thickness. After soldering, reduce the replacement to the same height as the other teeth and file the gear form on the top. Finish by filing it to the same width as the rest of the gear and polish both faces on emery paper.

Wheels that have all their teeth distorted have been subjected to an even, excessive load and, as a first measure, the wheel should be replaced with a new one with the same pattern of crossing-out. Do not leave it at this, there has to be a reason for the overload. If you have already discovered very stiff pivots or dirty train teeth, it is probable that this is the sole cause, but always suspect any bushed holes of having been drilled out of true centre and thus altering the meshing of the gears. Punching the outer edges of holes has the effect of altering the centre distances of the arbors as well as butchering the appearance of the plates. Check the centre distances by running the arbors concerned in a depthing tool and obtaining the distance at which smooth, even running occurs. You will be testing for easy running, no sudden stopping of the wheels when spun and no 'bounce back', any one of which will indicate that the distance is wrong. A punched-up hole can be drilled out with a size of drill that removes the punch marks as well (use your judgement, you cannot put huge holes in the plate); the hole can then be plugged and the depthing tool used to scribe the correct centre distance on the plug. File and polish down to plate level (see Chapter 3).

POCKETING OF PINIONS

Presumably because brass is relatively soft and grit tends to be pressed into it, the steel pinions are usually more worn than the wheels, which behave rather like grindstones. Each leaf of the pinion has a small pocket cut into it (Fig 12). This type of wear only begins to affect the working of the clock when the pockets are deep enough to enfold the wheel teeth, and any fall of dust onto the train provides material that jams between the top of the wheel tooth and the bottom of the pocket. If the clock is left in this state the speed at which wear takes place on the gear teeth and the pivots/pivot holes will accelerate. I mention this because, on occasion, clocks with badly worn pinions have had extra weight added to make them go and save the cost of replacing pinions or carrying out any work on the old ones. It is a false economy, which can only lead to tremendous damage to the train and pivots in the relatively near future.

If there is sufficient metal left on the worn pinion to retain a strong leaf after the pocket has been removed, the best solution is to soften the pinion and use a file to restore a full gear form to the leaf (Fig 13). The leaf will be much thinner but will operate well. This correction is fairly easy when the pinion stands alone, it is not easy when the pinion carries a wheel on its circumference, as it often does. The wheel will have to be removed and there is no certainty that the modified pinion will withstand the remounting of the wheel as described in Chapter 4.

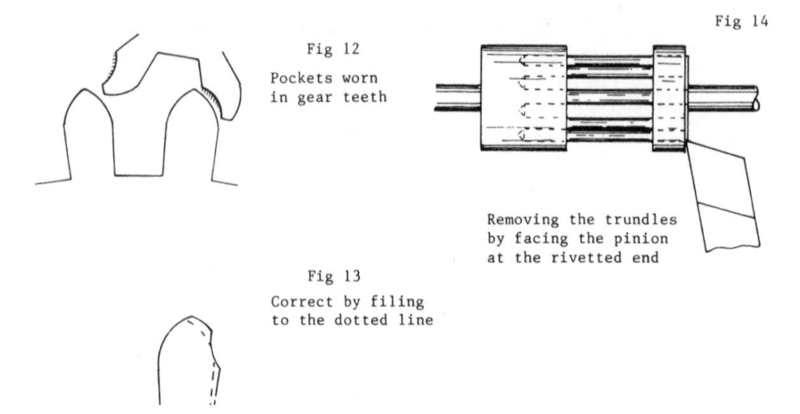

Pinions that cannot be restored, and those that would create more work in restoration than is justified, must be replaced with new parts.

WEAR ON WHEEL TEETH

Although worn pinion teeth are more common than worn wheels, there is no doubt that the latter do occur in normal service. However, since a major cause of worn wheel teeth is excessive load as a result of increasing the weight or fitting too strong a spring, carry out an examination of the drive when you see this type of wear. As a very general statement that is by no means always true, wear due to excessive pressure on the wheel teeth causes displacement of the brass sideways so that the face of the wheel has a raised portion around each worn area; wear caused by abrasion removes the metal by cutting and leaves hardly any raised metal around the area of wear.

I would not advise the filing of wheel teeth; it is simpler to turn the wheel around on its collet so that the other, previously unused, side of the tooth is presented to the pinion.

LANTERN PINIONS

The lantern pinion was used on cheaply produced clocks of the nineteenth and earlier twentieth centuries; it does not appear to have been used in mass-produced clocks since the 1960s. It also appears on turret clocks and a *few* high-quality eighteenth-century clocks.

It is not common for much wear to show on a lantern pinion and, since the train can accommodate wear up to the point that the pinion loses its accuracy of pitch, it is even more unusual for this pinion to give rise to a train failure. Correction is made by removing the old trundles (the bars of the lantern), and replacing them with pieces of pivot steel. Removal is achieved by either facing back the brass collar that supports them (Fig 14), or by cutting the middle of the old trundle with side cutters. In both cases the trundle is then slid out with narrow-nosed pliers, and the holes drilled with a drill of the same diameter as the trundle to remove burr and old dirt. The replacement trundles can be held by swaging the metal of the collar over their ends – you will have seen the original swaging on the face of the old collar – or by using an industrial adhesive such as Loctite, a cyano-acrylic.

If the pinion is so damaged as a result of a sudden shock to the train that it cannot be used again, turn the new one as a spool-shaped piece of brass so that both ends of the pinion – originally separate pieces – are joined by a mid-section of tubular form.

Rack-striking train and front-work

The tests for this train are, of course, the same as for the going train; but the following faults can also prevent the train from operating.

FAILURE OF WARNING

The warning is that movement of the train which, initiated by the lifting piece, puts the train ready to strike but held by the trigger effect of the warning piece. The lifting piece is the composite lever that extends one brass finger down to the minute (or cannon) wheel which carries a lifting pin, and the other finger raises the rack hook. The finger that lifts the rack hook has an extension at right angles that passes through a hole in the front plate and is able to interfere with a pin on the face of a train wheel. These are called the warning piece and warning wheel respectively.

Raise the lifting piece by hand. If the train runs and then fetches up on the warning piece (its proper action), there is still the possibility that in normal running the clock is not lifting the piece as far as you have lifted it by hand; so check this. If the train does not warn – making a limited rotation – when the lifting piece is raised, the fault is in the gathering pallet, the rack spring or the rack hook. If the train does not stop rotating after lifting the lifting piece, the fault lies in the position of the warning piece, or in the condition of the pin on the warning wheel.

Before stripping any of the front-work of the striking train, check the posts for looseness. The fault may simply be the result of the posts moving slightly or, in the case of a missing taper pin, a lever or wheel sliding backwards and forwards so that sometimes proper engagement is achieved and sometimes not.

MINUTE WHEEL

The minute wheel carries a pin that lifts the lifting piece, and over the years it can become worn or bent so that it does not lift the piece high enough to start the warning. Check by putting pressure on the great wheel so that the train is ready to turn – in the right direction – and turn the minute wheel. Check that the lifting piece makes full contact and is not in danger of slipping off the end of the pin. Make sure that it is raised sufficiently to free the rack hook from the rack teeth and that the gathering pallet turns until the warning wheel stops the train.

The first fault (not making full contact) can be the result of a distorted lower finger on the lifting piece, a bent pin on the minute wheel, or a very loose-fitting pipe and post to the lifting piece. When making this test, wiggle the lifting piece to make sure that a fault cannot be brought about by shaking of the movement.

Insufficient lift can be caused either by the lower arm of the lifting piece being twisted (turning the minute hand anti-clockwise will do this) or by it coming free of its soldering to the supporting pipe. If neither fault is evident, it is quite possible that lifting piece, rack or rack hook are not original to the movement or that some other repairer has modified them.

WARNING PIECE

It is possible for the lifting piece to grip the warning wheel, or for the wheel to press down on the flag so that pressure builds up on the minute wheel and stops the going train. This usually comes about when the flag on the end of the warning piece is incorrectly angled. If it is angled so that its face (where the pin on this wheel strikes) is not radial to the centre of the warning wheel, the pin will tend either to hold the warning piece in its raised condition or will oppose the lifting action, which in turn throws pressure onto the minute wheel. Beware too of pits worn in the working face of the flag, occasionally these will interfere with the latter's proper operation.

RACK HOOK (OR PAWL)

Two parts of the pawl are important apart from the fitting of the post that supports it, the part of the arm that is lifted by the warning piece, and the tooth that engages in the rack. Most longcase pawls are relatively slender in the main part of their length broadening out, scimitar fashion, towards the outer end. The slimness allows a certain amount of adjustment by bending so that movement of the warning piece will lift the tooth out of its position in the rack, and the broadening puts sufficient weight into the piece for smart operation. There should never be any need to add spring assistance onto a longcase drop, gravity should be quite sufficient.

The tooth of the drop is shaped so that it will ride up over the rack teeth as they are moved to the right by the gathering pallet. It should also have a slight slope on the opposite side to throw the rack tooth forward as the gathering pallet finishes moving the rack and comes clear of the teeth, at this point the pawl drops off the tip of one tooth and into the space between. If the pawl tooth falls onto the slope of the rack tooth without pushing the rack clear of the pallet there will be a backward shock as it hits the tooth slope, and the pallet will suffer additional pressure from the tooth it is driving. A slight slope on the pawl tooth makes for neat, precise movement of a rack that will suffer less wear.

The shape of the pawl tooth is such that the rack is left at the end of one tooth's movement, with the next tooth just within the arc of the pallet and slightly less than half its pitch from a line drawn between pallet arbor and rack mounting-post.

If you decide that a rack hook requires alteration to operate correctly, bear in mind that there are three faces that affect position – the face that the warning piece presses against and both faces of the tooth. It is safer to mark the line of the new surface and try it for effectiveness than to file away and then discover that your original idea of what needed to be done was wrong. It is rather like the problem of getting impulse and drop-off faces right on an escapement.

GATHERING PALLET

The tooth of a gathering pallet can wear so that it no longer throws the rack far enough forward at each revolution for the pawl to drop into position. This is usually apparent on one or two places on the rack. In other words the clock will strike correctly at certain times of the day, but add in a few extra strikes at others. It may be so bad that it continues striking until the train is run down.

Correction can sometimes be made by filing the righthand-facing surface of the pawl so that the point moves to the left and falls into the rack tooth sooner. If this cannot be done, the only remedy is to add a little metal to the tooth of the pallet. Remove most of the existing tooth so that a fairly large seat can be obtained for the new one, and silver solder a piece of silver steel there. Plunge into water as soon as the joint is made, or pour water on, so that the new metal is hardened. Now temper blue and file to shape with a saw-sharpening file. Polish on Arkansas stone to finish and do not leave a sharp edge on the tooth.

A striking train that starts to move at the warning and then refuses to complete the strike, is having to work too hard at the beginning of its operation. Almost certainly you will find that the hammer is left on the pin-wheel at the beginning of the strike so that the train has no chance of turning before being called upon to lift the hammer, or the hammer spring is too hard, or has made a notch for itself in the hammer lever. The first can usually be catered for by taking the gathering pallet off the arbor and rotating it until the hammer comes off the lifting pin just before the gathering pallet stops the train.

Clocks that are required to make the first strike precisely on the hour almost always leave the hammer resting on the lifting pins at the end of the strike, but they are then arranged so that the lifting lever of the hammer arbor is tangential to the pins when resting there, which ensures that the train is not loaded when the strike begins (Fig 15).

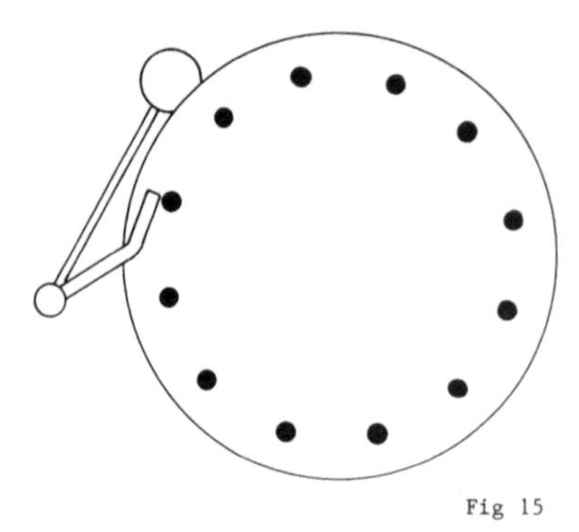

Fig 15

Hammer and hammer tail

RACK AND SNAIL

The striking train that progresses through the lift and warning but varies in its number of strikes, is almost always failing at gathering pallet, rack or snail. The pallet has just been dealt with, let us now test the rack and snail.

Advance the hour hand and snail to eleven o'clock. Few longcase clocks have the snail on a star wheel, but if this is the case then remember to synchronise the snail with the hour hand after all is done. Run the train by hand, going through each strike, and check whether the three positions – 11, 12 and 1 – have the correct relationship to one another. By that I mean that, if there is incorrect striking, the difference between the hours of 11, 12 and 1 o'clock remains one strike and eleven strikes respectively. If this is the case, then proceed to 'pitch and lever-arm faults' (below). If, however, you are getting variations at this point it is probable that the clock has been dismantled by someone not familiar with striking problems, or that the rack, snail or gathering pallet are not original, but 'spare parts' that have been put in optimistically.

Simple faulty adjustment will allow the tail of the rack to fall onto one end or other of the snail segments, so that wear causes one strike more or less than is proper. It can also result in the rack tail falling onto the radial drop between 12 and 1, giving any number of strikes between 11 and 2 and quite possibly stopping the clock at about five minutes past the hour. Re-adjustment is often merely a matter of disengaging the hour wheel and the cannon wheel, and advancing or retarding the meshing point. If the hour hand is pinned to its pipe, you may find that correcting the strike will leave it pointing to one side or other of the hour division. All that can be done is to check that the hand has not been bent at some time, thus causing misalignment, or to re-pin the hand on its pipe.

Faults that result from using a 'spare part', call for remaking either the snail or the rack tail. The latter is simpler. If the clock is not giving enough strikes, the tail needs to be shortened; if giving too many, lengthen it. In both cases use the old tail for establishing the right radius for the pin that registers on the snail, by drilling a small hole at the estimated position and inserting a taper pin. Try the train by hand to prove the position, and then either replace the tail completely or put another register pin in the new position.

Pitch and lever-arm faults Over a period of years the pitch of the rack teeth varies as a result of wear or mishandling. This will evidence itself in a tendency for the strike mechanism to fail to gather properly at certain times. This pattern of failure will be repeatable, for example a clock may often fail on a particular strike, say seven o'clock, but never anywhere else, or the clock may stop every time it is required to strike five o'clock.

The only cure is to mark the pitch of the rack teeth, using a pair of dividers. Set the dividers to a distance equal to two pitches (Fig 16). Using a file, ensure that the first two (left-hand) teeth are of equal pitch. Place one leg of the dividers at the bottom of the radial part of the first tooth and mark the arc for the third tooth, move the leg to the second tooth and mark the fourth, and carry on in this manner until all are marked. Now file the radial faces to the marks scribed by the

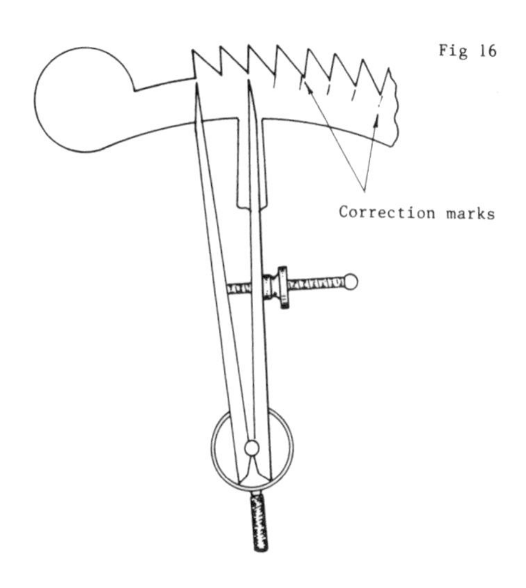

Checking the pitch of rack teeth

dividers. If one of the teeth cannot be filed back to its mark, insert a piece of steel, solder it and then file. Use a saw to put a slot at the base of the tooth before soldering, and then use this to hold the new piece.

Snail wear Similar faults to those resulting from a damaged rack will appear if the snail's segments are worn. You can probably alter the position that the rack pin falls onto and find an unworn space at each segment, but if this has already been done you will have to correct the snail itself. Make sure that the lowest segment – twelve o'clock – can be filed without fouling the pipe that it rides on. Sharpen the face of the rack pin by filing a sloping front to it, the edge coming at the point that is expected to touch the snail; break the soldered joint between the rack tail and its pipe and open up the angle slightly. If this opening up has placed the rack pin slightly below the snail surface when the rack is positioned for the twelve o'clock strike, use it to scribe an arc on the face of the snail in the twelve o'clock segment. If the rack pin has not moved to scribe a line a little lower than the old worn surface at twelve, open the angle a little more. Now scribe an arc on each segment with the rack hook resting in the appropriate rack space. The snail is usually fastened to the hour wheel with two steel rivets, which will need to be driven out after marking the segments and before filing.

Locking-plate striking

This is also referred to as count-plate or count-wheel striking, although strictly speaking there are slightly different definitions for each. All perform the same duty, that of establishing a sequence of strike patterns. An ordinary clock has the pattern one o'clock, two, three etc; a half-hour striking clock interposes a single strike between each hour; a Comptoise repeats the hour strikes after a couple of minutes; a Roman strike uses different bells for I, V and X – and even this is not the limit of striking variations. However, what is common to all locking-plate systems is that the pattern is unvarying in sequence for a given clock. It cannot be asked to 'repeat', ie strike the time upon request, and the hands must be arranged to coincide with the hours being struck when the clock is started. The clock will not automatically strike the right time as a rack strike does.

The system is often used to control chiming on a clock that has rack strike. In this instance the sequence of the quarters can be corrected automatically, after an alteration of the hands, by stopping the sequence of chimes just after the three-quarters and ensuring that it cannot start again until the minute hand comes up to the 12 position. (This does not invalidate the previous statement regarding the striking of the hours.)

Since any failure to strike at the hour will result in hands and bell being out of agreement with one another, occasional faults will be made apparent by the striking mechanism. A rack strike may leave some doubt as to whether the hour has just been struck, or whether one has simply not noticed it, a locking-plate strike is not so courteous – everyone will know as soon as it is time for the next striking.

Lifting and warning are tested in the same manner as for a rack strike.

HOOP WHEEL OR STOP WHEEL

This is the position where the train is usually stopped and held ready for the next sequence; failure will be due either to wear (or bad workmanship) at the radial face of the lever that stops the wheel, or because the relationship between this lever and the locking-plate hook does not allow the proper positioning of the stop lever or stopping piece. Check the positions of these two levers, the contact face and, in the case of a stop wheel that carries a pin, the length and condition of the pin. American clocks are sometimes stopped by a lever catching the fly, as are some modern British and European clocks. Since it is important that the stop position does not shift about, the fly does not follow normal practice and slip on its arbor, it is a positive fixture. However, if the stop is on the fly arbor and not the actual 'fan', normal practice is followed.

LOCKING PLATE

This part, that gives the mechanism its name, can be either a plate with cut-outs around its periphery, a hoop with cut-outs, a plate with pins around the periphery arranged radially or parallel with the axis, or a toothed wheel with a series of deeper cuts at the hour or hour and half-hour positions (see Chapter 10). The slots or pins are simply marking the end of a free-running period of the striking train, which runs and strikes until the end of the period is reached, whereupon the stopping piece halts it. Since the number of strikes in a twelve-hour period is 78, or 90 when half-hours too are struck, the locking plate is divided into 78 or 90. The end of the period is 'sensed' by the locking lever dropping into a space or resting on a pin.

Most failures of locking are due to the locking piece catching on one side or other of the space that it falls into. Allowance must be made for any movement that the locking plate can make before the locking piece has lifted clear. This lever also holds the stopping piece clear when the clock is required to strike. The relationship between the two must be satisfactory, otherwise the stopping piece will fail.

Adjustment of German, British and American clocks can be made by bending the locking-hook lever-arm or the blade itself. In many cheaper clocks these levers are made of wire which can be adjusted by means of the tool shown in Fig 17, otherwise use smooth-jawed pliers. French clocks often have a friction-tight joint between the locking piece and its arbor.

Since the positions of the cut-outs on the locking plate do not have to be precise and they should not suffer wear, it is unusual for these to be wrong. Check everything else before deciding that there is an error in the

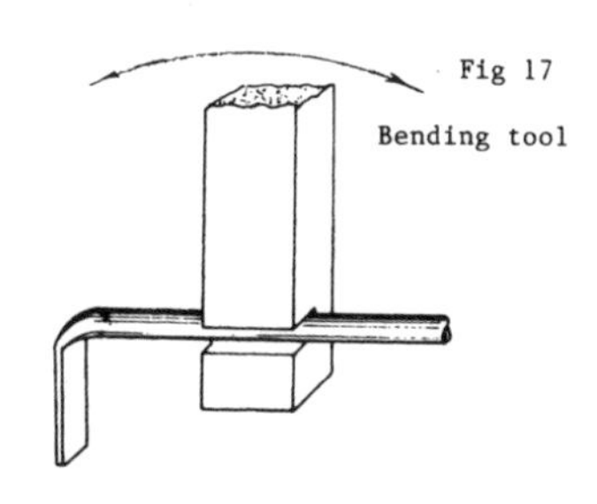

Fig 17
Bending tool

plate and, if there is no evidence of any 'bodging', check again.

CHIME BARREL

If the chime is giving the wrong notes for the quarters, ie if notes from one quarter are not being struck until the beginning of the next, the barrel has changed its position relative to the chime count plate. There are many arrangements of chime, but there is usually some means of turning the barrel without moving this plate. Frequently it is done by having barrel, idler gears between barrel and count plate or the count plate itself fastened to their respective arbors by means of a grubscrew. Loosening the screw, or screws, enables one to turn the barrel forwards or backwards. Make sure that you are familiar with the chime tune before embarking on this.

When dealing with a strike that has automatic correction remember that, during the time that the clock is correcting itself, the chime will not work. Overlooking this simple fact can lead to much unnecessary work.

Movement plates

Squareness of the movement plates is fairly obvious and can be determined without stripping, although correction will result in the train being removed. What is not always obvious is that plates can distort when the weights are attached, or when the seat-bolts or hooks are screwed tight. If a clock movement shows all the signs of having a tight train when mounted in the case, but no tight pivots or meshing pairs can be found, suspect distortion of the plates – very old clocks with slender pillars that have been drilled and tapped in the centre are very prone to distortion. The seat board should be checked for flatness and the seat-hooks registering on the pillars close to the shoulders replaced by new ones; insert the old screws, however, but not tightly. Sometimes this fault can be traced to the use of large washers under the seat board that prevent the nuts sinking into the wood when over-tightened.

CORRECTING A DAMAGED SET OF PLATES

Dropping a clock movement is a not infrequent reason for distorted plates and correction must depend upon what has actually happened. If the pillars are slender and the plates stout, it may well be that the former have bent but the seating into the back plate is still secure. Lay the plate flat on the bench and slip a piece of brass tube over the pillar, using this to ease the pillar square to the plate once more (Fig 18). Bear in mind that the pillar is a casting, probably with a fairly coarse grain, and that it is old; if it breaks you will have to turn another.

If the pillar itself has not bent (Fig 19), it is extremely unlikely that it can be moved back without damaging the riveted seating into the back plate. Try it by all means, but be prepared to carry out the following repair.

Remove the pillars from the plate – those that are not square, or those that show light under one side of the shoulder. Flatten the plate with a polished planishing hammer using a piece of brass plate as an anvil. Support the pillar in the lathe with a steady (Fig 20), and face the shoulder back slightly – 0.25mm (0.01in); do not worry if the origional rivet will not run true, but make sure that the new diameter at the shoulder is smaller than the original. In other words, undercut the old diameter of the rivet. Gently use a cutting broach to make the pillar rivet enter the hole in the back plate; the broach enters from the outside surface of the plate so that the rivet will be 'made' in a slightly tapered hole. Ensure that the inside surface is flat by placing a straight edge on it and holding up to the light. Before riveting the pillar, make a support for the other end (Fig 21) in the form of a piece of steel drilled to accept the pinned end of the pillar and then faced so that the shoulder of the pillar will lie on a smooth surface. Rivet with a ball pein hammer or the 'peck' on a planishing hammer, beginning with light taps around the protruding edge of the rivet and finishing in the centre. File flat and polish. All filing should be done in the direction of the long axis of the plate (see Chapter 3).

The plates should now be square, but the distance between front and back plates has decreased by the amount that was removed from the shoulder of the pillars, and if only one pillar had to be re-riveted the plates will not be parallel. If the latter is the case, machine a thin washer to make up the amount of metal removed and fit it between the shoulder of the pinned end and the plate. The pin hole will very probably have to be opened out with a cutting broach before the movement can be pinned together again (Fig 22).

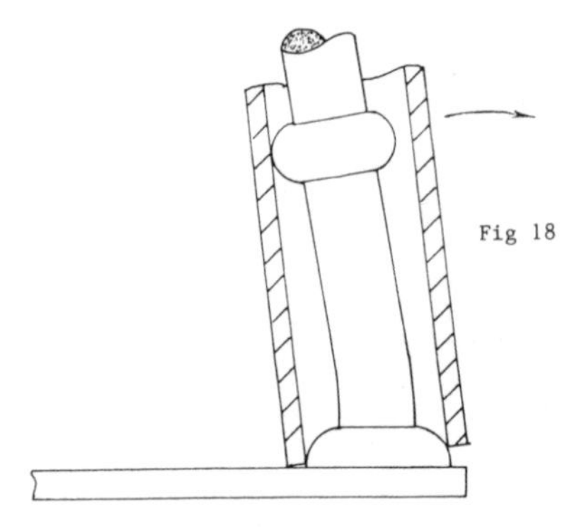

Fig 18

Straightening a pillar

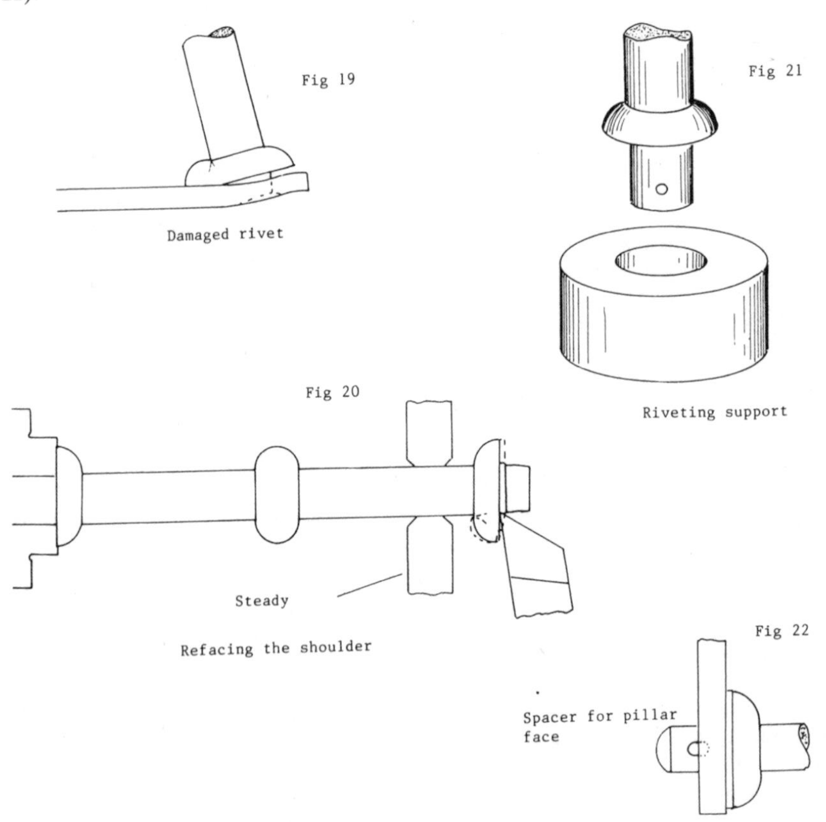

Fig 19

Damaged rivet

Fig 21

Riveting support

Fig 20

Refacing the shoulder

Fig 22

If all pillars have been machined – do make them all the same length in this event – you can either make up a complete set of washers as above, or file back the shoulders of the arbors until they all have movement between the new position of the plates. This is not always as tedious as it sounds, because if the pivots have been re-polished several times in their life the shoulders will probably have been filed back anyway, and if new pivots are needed it is no problem to face back these shoulders when re-pivoting.

ADDITIONAL FAULTS IN SPRING-DRIVEN CLOCKS

Very many of the faults covered already are common to both spring- and weight-driven clocks, those that follow are specific to spring-driven ones.

Open spring

An open spring is a feature of many American antique clocks and modern alarm clocks of all nationalities. Its major failing lies in the probability that, in an elderly example, the spring will interfere with the rest of the movement as it runs down because the stop pin that should restrain the spring gradually gives way over the years. Otherwise, it does not suffer from the problems attendant on the enclosed spring – difficulty of lubrication, and friction between coil and coil. Extremely variable timekeeping on a day-by-day basis may well be caused by the spring twisting as it expands, and rubbing against the plates or even the wheels. There is no other remedy for such a distorted spring than to replace it. A tendency to stop short of its full term (twenty-four hours or seven days) indicates that the spring is either 'tired' (see Chapter 12), or that it is being allowed to interfere with the going train.

The stop pin mentioned above, that stops the spring interfering with either the going train or the wheels of its own train, appears in many clocks. It is usually fairly stout and riveted into one of the plates, sometimes one of the movement pillars carries out the same duty. Since the pin protects the rest of the clock, the spring can continue working after the point when it would otherwise have interfered, and the outer coil comes into contact with the stop at a point where no sliding takes place between them. As a result the effective centre of the spring is shifted and the spring continues to open in a normal fashion without undue friction between coil and coil. The centre of an open spring is not in a stable position; it moves as the spring opens or is wound up.

There are two types of outer anchor for the open spring – the riveted loop and the open loop. The latter enables the repairer to feed a new spring into the assembled movement from the side and then slip the open loop over the anchor post (which is also one of the movement pillars). A spring with a riveted loop can only be fitted by taking off the front plate. A considerable amount of energy remains within an unwound open spring and this is only controlled by the plates, therefore, when dismantling an unbroken spring, wind it up first if possible and slip on a semicircular clip, then allow the spring to unwind into the clip until it is gripped firmly. Clips are available from suppliers.

The strength of a new spring is very important, for the wheels of these clocks are relatively thin and will wear badly if oversprung. It is not sufficient to measure the old spring and take that as the datum for the new one. If a train has been wearing gradually, the teeth of the wheel will show pockets but very little fraze at the side. A clock with too powerful a spring produces worn wheels with squeezed out fraze or burr. Fit a lighter spring. These clocks, although cheap, have often run for the best part of a century and have at least as much life left in them if treated well.

Going barrels

This is the common method of driving a British, French or German antique spring clock and very many good-quality modern ones. Faults due to worn pivots are discovered in the same way as on other arbors of the clock. However, the working pivot is not fitted in the movement plate, that is only used for winding, it is in the end of the barrel. Any faults of mesh due to worn pivots must be corrected by removing the barrel and bushing its toothed end.

A broken or worn tooth can rarely be repaired by inserting new metal because the root of the tooth is so close to the outer diameter of the barrel and because the bore of the barrel passes beneath the tooth leaving a thin disc at the end that is usually between

1mm (0.04in) and 1.5mm (0.06in) thick. It has been the practice of some repairers to 'peg' a broken barrel tooth with steel pins drilled into the root of the missing tooth. This is bad practice since in most cases the working pressure is raised because of the reduced gear-tooth area and the poor working of steel teeth on steel pinion. Such a repair will only add to the cost of the next repair.

If a spring is badly distorted it can rub hard on the ends of the inside of the barrel, producing metal shavings and making a very nasty mess. The spring can often be reshaped by careful stretching out on a clean surface and judicious twisting to persuade the coils to lie within the plane of winding. It is a skilled job and calls for fairly strong hands. The alternative is to buy a new spring.

Insertion and removal of springs can be carried out using a spring winder, or by hand. If you use the hand method, make sure that the spring is induced to come out rather than simply hauling it out. Chapter 12 gives a description of this operation, and if you have a spring winder there will be instructions that relate to that particular model. When inserted, the spring should not rub on either cap or the base of the barrel (Fig 23).

The best lubricant for springs is silicone grease with molybdenum disulphide; it will work for decades without changing its characteristics. Unfortunately, if a future repairer adds oil to the barrel, a sticky mess results; therefore until we can all agree to use grease in going barrels and fusee barrels, I would advise limiting its use to your own clocks or at least those that you know will return to you for servicing.

If the anchorage of the spring to the inside of the barrel becomes detached, it will need to be re-riveted in place or a new one made. The anchorage is usually riveted in the barrel wall and this gradually works loose. To make it firm again, grip a steel bar that you have drilled a hole in of larger diameter than the anchor in the vice, and use it as an anvil (Fig 24). Slip the barrel over until the anchor drops into the hole, then use a light hammer to tap the metal surrounding the rivet until all is snug again.

German barrels sometimes have a tag partly pressed from the wall of the barrel to act as an anchorage. If this breaks off or starts to tear, make sure that the inside is clear of all remnants and then fit a new steel anchor further around the diameter of the barrel and the same distance from the end as the original anchorage. This anchor is similar to that normally fitted in going barrels and looks like a flat sailor-hat. Its largest diameter must enter the largest part of the hole in the spring and then, as the spring is drawn onto the anchor, prevent it coming off again. The form of the anchor can be seen in Fig 24. Of the three diameters that make it up, the middle one is the one that has to fit into the anchor hole in the spring, and it must be long enough to allow the thickness of the spring to settle neatly behind the largest diameter; the smallest diameter forms the rivet into the wall of the barrel and should protrude sufficiently to grip the wall firmly when the rivet is ex-

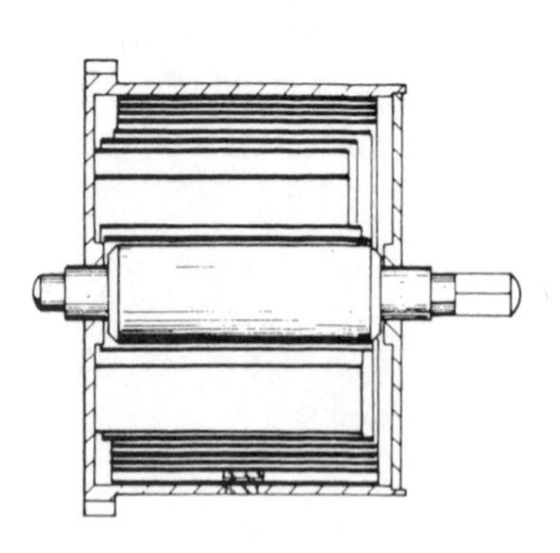

Fig 23

Cross-section of going barrel

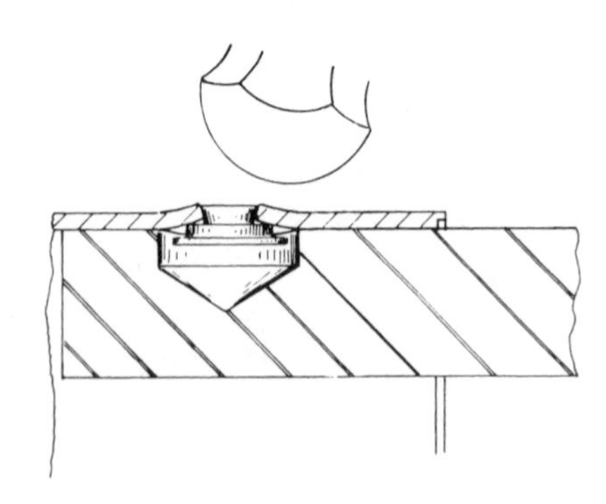

Fig 24

Tightening spring anchor

panded into it. The material used is steel – silver steel for most purposes, but on very strong springs a stronger anchor may be obtained by turning down a high-tensile bolt or Allan cap-head screw. Do not harden.

Winding arbors carry the second (inner) anchor. A spring that will not wind up, but does not show any sign of having broken, has probably lost contact with this anchor particularly if the clock was allowed to wind right down and was then left for months or years. Broken springs either go during winding or result in a stoppage part-way through the going period and then, of course, will not wind up; an anchor cannot merely lose contact when partly wound up and running. Either fault will necessitate dismantling the clock. Springs that have simply lost contact can be corrected by using a pair of thin-nosed pliers to close the inner coil tightly onto the arbor.

Broken springs can be repaired if the very end of the spring has broken off or the anchor hole has ripped open. Soften the spring by heating to red heat and then cooling slowly, the colour produced by this process should not advance further along the spring than the broken piece indicates was the original state. If the original piece of soft spring was 30mm (1.2in) for instance, the new end should be no more. Clamp the spring between a flat piece of metal and a board leaving the position for the new anchor hole exposed, and drill a hole. If you are intending to do this on a pillar drill, clamp to the table or risk damage to your hands if the drill 'grabs' the spring when the point breaks through. Not all springs can be satisfactorily re-drilled because the metal adjacent to the annealed area tends to become brittle – old American springs are particularly prone to this failure. You will have to decide for yourself whether to carry out the repair or to buy a new spring. In any case, if the original hole shows signs of having been filed and the spring is not a very old one, it is likely that it has already been re-drilled at least once and the spring will be too short for you to carry this out again.

Inspection of the spring when the cap is removed from the barrel (but without extracting the spring) will clearly show any break. The spring ought to be inspected when the clock is dismantled because the anchor hole may well have torn and be close to complete failure. This results in the spring end being lifted out of the plane of the rest of the spring (Fig 25), when it can be seen whether it is at the outer or the inner end.

FUSEE AND BARREL

The fusee is a device for obtaining a less variable torque at the great wheel than would otherwise result from a spring-drive. Very little can go wrong with it other than failure of the clicks and click springs and, of course, pivots and pivot holes. There are very often two clicks since maintaining power is included in the quality of clock that usually has fusees, and the maintaining power has a ratchet and external click or pawl.

SETTING SPRINGS

The spring in a fusee barrel must not wind down completely when all the cord has wound off the fusee, otherwise the torque at the great wheel would vary considerably from the beginning to the end of the going period. The amount of energy that is left in the barrel when the fusee has completed its run is called the 'set', and very commonly is a simple matter of adjusting the tension on the cord between fusee and barrel to be nil when the former has no cord wound onto it, and then using the barrel ratchet or click to tension the cord with one half turn of the barrel arbor. This is fine for clocks with no pretensions to accurate timekeeping, but to get the best performance from a spring you should follow the more complex method described in Chapter 12. Going barrels with 'stop-work' to govern the portion of spring used, also need to be set when assembled, this is also described in Chapter 12.

TIRED SPRINGS

Any spring contained in a barrel should have

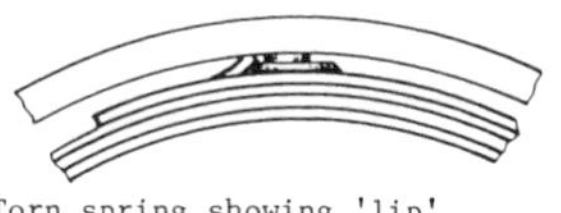
Torn spring showing 'lip'

Undamaged spring end

Fig 25

no more than one or two coils crossing the space between the arbor and the unwound coils lying against the diameter of the barrel – I am talking about unwound springs of course. The energy of the spring can only be transferred to the movement by the coils that move from the wound condition to the unwound, and any coil that does not unwind to lie against the inside of the other coils is not working. It is 'tired' and the clock will show it by not running for its full term. The spring should be replaced. Some German mantel clocks make use of a spring that does not need to make use of its total possible energy and may look tired when new.

Suspensions and pendulums

SILK-THREAD SUSPENSION

This takes the form of a simple thread attached at two points, with the hook of the pendulum rod caught on the loop between. Some are adjustable from the front by a rod that protrudes over the top of the dial, others are adjusted by turning a small turned knob at the back. The pendulums are very light as a rule.

A broken thread may be replaced by a light mercerised thread or glazed cotton; button thread is usually too thick. Sylko machine-twist of 40 gauge is a very good substitute for silk thread and a light rub with beeswax would be useful. Silk thread is the ideal, but it will make little or no difference to the timekeeping. All threads will vary with the weather and temperature and clocks of this type are not good timekeepers.

BROCOT SUSPENSION

This is usually adjustable from the front. The spring cannot be repaired, it must be replaced; the main points to consider for the replacement are:

1 Shape and size of top block
2 Width and thickness of spring
3 Size of bottom block

The overall length of the spring is not too important as long as it does not affect the operation of the crutch or seriously alter the effective length of the pendulum.

The spring should be clasped closely by the brass slider in the Brocot back-cock, but not so tightly as to tear the spring when adjusting its effective length or moving it to hang plumb. This is one type of suspension which rarely allows the weight of the pendulum to pull the suspension truly vertical, and it is quite usual to have to give a *little* tug to the bob to set it plumb. Tightness of the block can be altered by using a thin brass wedge to open the spring slot to loosen it, or to open the slots on either side to tighten it.

KNIFE-EDGE SUSPENSION

There are two forms of knife-edge suspension besides the wedge-shaped blade on a brass or steel bed. One is the form used in the Kee-Less clock of the 1920s – a gravity clock with a small compound pendulum. Here, the suspension is a simple arbor with plain cylindrical pivots resting in pivot holes that are several times larger in diameter so that the pivot rocks on the larger curve of the hole.

More common is the Black Forest loop and staple. Here, refusal of the clock to 'go', even when movement and pendulum are in good condition, can often be traced to the loops wearing into the staple so that the contact between the two is no longer point contact, but face on face. Either use a new staple made of pivot wire or, if a large proportion of the metal is unworn, adjust the width of the loops to bear on an unworn place.

PENDULUMS

In general there are few faults with the pendulums of mantel clocks and others that use short rods. The one that will keep occurring concerns the simple disc used on some Black Forest clocks, which is held in position by a bowed spring or by a slight distortion of the disc itself. Unexpected changes in rate can often be traced to this disc's slipping. Put more 'set' into spring or disc.

Motion work

Hands that do not move when the clock is going, and unexplained gains or losses in the course of one hour, can be put down to faulty motion work including the friction drive of which there are two main types: the French, that causes friction between the centre arbor and the inside of the minute pipe, and the sprung type. The first can be cured quite simply by tapping the side of the pipe lightly with a hammer. Very often the pipe has had two flats cut deeply, one on each side to form two straps connecting the top and bottom of the

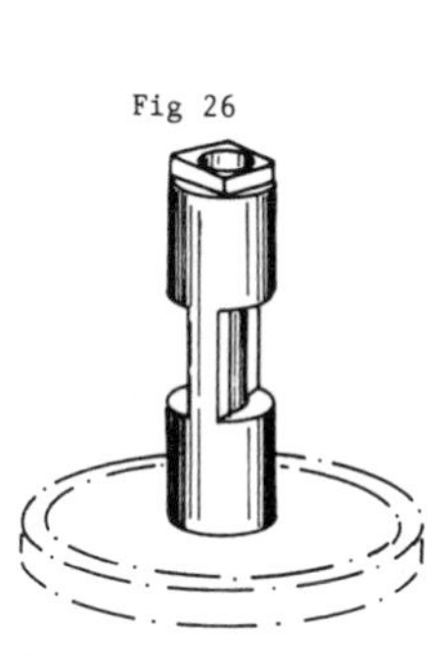

Fig 26

French minute pipe friction drive

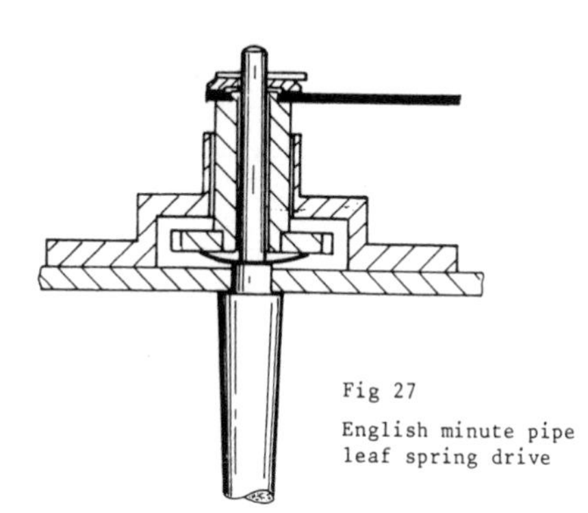

Fig 27

English minute pipe leaf spring drive

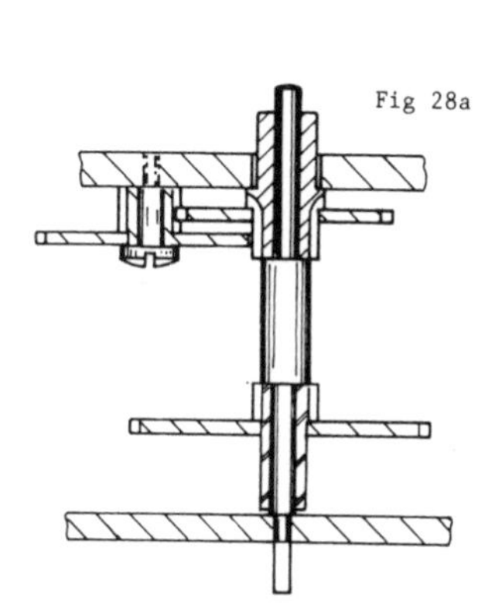

Fig 28a

Minute pipe and motion work of modern carriage time-piece friction drive

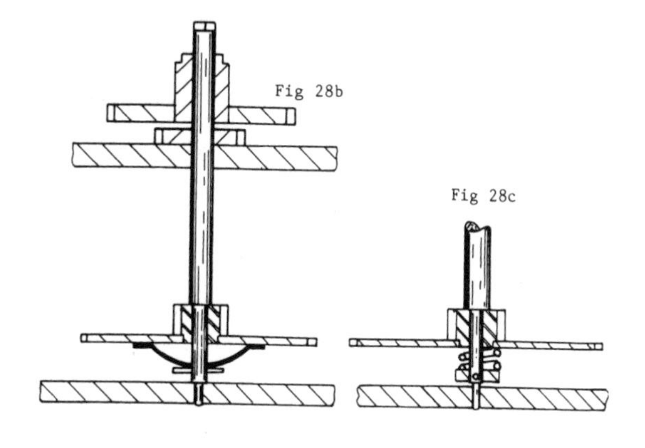

Fig 28b

Fig 28c

Minute pipe and motion work typical of Black Forest and American clocks

pipe (Fig 26). Tapping these inwards is very easy, do not use too much force.

The sprung version is used in British, German and American clocks, and is either a leaf spring or a coil spring. The leaf spring is trapped between a shoulder (sometimes squared to fit a similar hole in the spring) and a wheel (Fig 27). The amount of drive depends on the depth of the bow in the leaf and the amount that the wheel is pressed down onto it when the taper pin is inserted. Correction is a simple matter of putting a little more bow on the spring or, if it is cracked, making a new one. It is only made of brass sheet beaten out and filed and drilled. The position of the spring may be between the cannon wheel and the centre arbor, between the arbor and the brass pinion that drives the hour wheel of a three-wheel clock, or between any convenient wheel and arbor within the plates.

The coil spring is frequently found in American clocks, where it fits over the centre arbor and bears against the faces of the cannon wheel and the centre wheel, lying between them (Fig 28). Often made of brass, any fault is usually due to grease or oil. Simply clean with petrol or white spirit. The coil spring is nearly always limited to clocks that have the motion work between the plates, and cleaning is sufficient to put a stop to the slipping.

FAULT FINDING CHART

This chart plots successive tests and corrections. The first column names the part or type of clock being tested, the second column shows a numbered test, the third shows typical results and the last suggests the corrections that should be made. The numbers in the third and last columns give the next test to be applied. For instance, test 7 (carriage clocks) examines the balance wheel: if it rocks the platform should be removed and renewed, and the next test is 17, to see if the rest of the movement is clean. If the platform does not rock in test 7, proceed to test 8 and slacken the base screws. If this results in the clock working, proceed to 69.

Please note that the first column introduces different types of clock and consequently some of the positions in the third column carry a series of numbers, each one relating to a different clock.

Clock type, train or part	Tests	Results	Correction
All mechanical clocks	Wind & check beat	Clock goes: 78 Clock does not go: 1 Goes for a few minutes: 3 Goes for one hour: 3 Goes for several hours: 1 Goes irregularly: 3	
Weight-driven l/case & Vienna l/case	1 Open case doors	Goes Does not go: 2	Correct fall of weight: 78
	2 Remove hood	Goes Does not go: 3	Correct interference with mov'm't: 78
Above & wall, bracket, shelf, mantel clocks, spring or weight	3 Remove hands	Goes Does not go: 4	Correct hands or leafspring: 78
	4 Remove dial	Goes does not go: 5	Look for loose frontwork or interference with hand pipes: 68
	5 Remove pendulum	Goes Does not go: 6, 7, 10, 14, 15 or 16	Correct suspension or pendulum: 68
All striking clocks	6 Move hands back from their stopped position four or five minutes	Goes Does not go: 7	Test striking: 51 or 56
Carriage clocks	7 Gently rock from side to side	Balance wheel rocks No rock: 8	Remove mov'm't and renew platform: 17
	8 Remove base plate & slacken *two* holding screws	Goes: 69 Does not go: 9	
All platform escape't clocks	9 Remove from case, check beat	Goes when in beat: 69 Does not: 17	
400-day clocks (anniversary)	10 Check level of base	Goes: 69 Does not go: 11	
	11 Check beat	Goes when in beat: 69 Does not go: 12	
	12 Very slightly slacken the mounting screws of mov'm't	Goes: 69 Does not go: 13	
	13 Check rating adjustment	Clock will not give correct time-keeping rate Rate correct, clock stops: 17	Renew suspension or change pendulum characteristics: 10
Maltese & all clocks with mov'm't on door	14 Open & support door, put in beat	Goes: 69 Does not go: 17	
Banjo clocks	15 Check verticality of front	Goes: 69 Does not go: 17	
Four-glass, crystal, clocks supported between columns	16 Check fixing screws of mov'm't to frame	Goes when loose, Does not go, loose or rigid: 17	Attend to mounting: 69
	17 Dismantle mov'm't: 18		

Clock type, train or part	Tests	Results	Correction
All mechanical clocks	18 Inspect for dirt and gummy oil	Goes after cleaning: 19	
		Does not go: 32	
All trains	19 Assemble & shake mov'm't	All arbors rattle: 20	
		Some do not	Increase clearances: 20
	20 Turn great wheel by hand	Escape wheel makes part turn under pallets	Correct pitch &/or eccentricity: 23
		Escape wheel turns but with difficulty: 21	
		Fly turns: 22	
		Fly does not turn: 22	
		All wheels turn: 21 & 22	
Going train	21 Remove escape pallets or platform, apply light pressure to great wheel, fusee or barrel	Escape wheel turns smoothly: 23, 26, 28 or 31	
		Does not: 32	
Striking train or chiming train	22 Remove frontwork, apply light pressure to great wheel etc	Fly turns smoothly: 45 & 56	
		Does not: 32	
Recoil-escape't l/case, bracket, wall, shelf, mantel clocks	23 Replace pallets, turn escape wheel	Pallets have an even drop around the full turn of wheel: 24	
		Pallets do not drop evenly	Correct pitch of wheel: 24
	24 Examine pallets for well-defined grooves or pits	Drop exceeds 10% of pitch	Re-face pallets: 25
		Drop does not exceed above: 37	
		One pallet has much more drop than the other	Re-face pallets: 25
	25 Measure arc of crutch to let teeth escape	Arc approx 3° about centre (arc is 1/10 length): 37	
		Arc much less	Re-face or adjust pallets: 33
		Pallet arbor lifts at each beat	Re-bush or re-pivot: 33
Graham dead-beat escape't l/case, bracket, Vienna, 400-day, small regulators	26 Replace pallets, turn escape wheel, move crutch by hand	Pallets have an even drop around the full turn of wheel 27:	
		Pallets do not drop evenly	Correct pitch of wheel: 27
	27 Examine pallets for well-defined grooves or pits	Drop exceeds 10% of pitch	Re-face & adjust pallets: 32
		Drop does not exceed above: 37	
		One pallet has much more drop than the other	Re-face & adjust pallets: 32
		Wheel drops directly on to impulse face	Adjust pallets: 32
		Pallet arbor lifts at each beat	Re-bush or re-pivot: 32
Brocot dead-beat French, German & some American mantel	28 Replace pallets, turn escape wheel, move crutch by hand	Pallets have an even drop around the full turn of wheel: 29	
		Pallets do not drop evenly	Correct pitch of wheel: 29
	29 Examine pallets for wear	Pallets worn	Replace pallets: 30
		Pallets not worn: 30	
	30 Check depth of pallets into wheel	No locking	Re-bush & new pallets: 34
		Too much lock	Re-bush & new pallets: 34
		Pallet arbor lifts at each beat: 33	Re-bush or re-pivot: 34
Platform escape't	31 Remove platform from mov'm't & gently operate escape-wheel pinion with one finger	Escape wheel does not rotate when balance does	New platform escapement: 32
		Escape wheel rotates when balance does: 37	
Pivots & pivot holes	32 Examine holes for ovality	Holes oval	Re-bush: 33
		Holes round: 33	
	33 Examine pivots for grooves	Pivots worn	Re-pivot: 34
		Pivots not worn: 34	

Clock type, train or part	Tests	Results	Correction
	34 Examine gears for meshing at pitch circle	Centres too far apart, butting	Plug & re-bush: 35
		Centres too close, rumble	Plug & re-bush: 35
		Centres correct: 36	
	35 Examine gears for uneven running	Pivots bent	Straighten or re-pivot: 36
		Pivots not bent: 36	
Gear teeth	36 Examine teeth for wear	Teeth worn or bent	Reverse wheels, file pinions or replace: 37
		Teeth not worn: 37	
Entire train, going, striking or chiming	37 Assemble train and weight or spring barrels	Going train operates less than a few minutes	Lubricate pivots: 62
		Striking or chiming train does not operate	Lubricate pivots: 62
		Going operates for only a large part of its nominal wind: 38 or 42	
		Train goes: 45	
	38 Measure drop of weight possible	Weight does not make use of whole drop	Extend cord or chain: 39
	39 Check cord or chain does not foul move'm't or seat board	Cord or chain catches	Re-position seat board, cord or chain: 40
	40 Check chain does not jam in chain-wheel	Chain has opened out	New chain: 41 or 62
		All free, clock goes: 62	
Front-work	41 Does front-work hold the minute wheel back?	Yes	
		Clock does not go: 45 or 56	
Spring barrel or going barrel	42 Remove barrel cap & inspect spring	Spring dry, possibly dirty	Extract spring, clean & lubricate: 43
		Too much spring	Shorten or change spring: 43
		Too little spring	Change spring: 43
		Metal dust and shavings in barrel	Spring too wide, distorted; change spring: 43 or eccentrically anchored; correct hole: 43
	43 Is cap loose or bores over-size?	Cap comes out in use	Beat the edge of cap to make larger diameter: 44
		Barrel ring runs eccentrically. *Note:* bores of fusee barrel not important	Bush barrel ring: 45
	44 Does fusee barrel give sufficient torque at end of wind?	Clock does not run for full nominal wind	Turn ratchet on barrel a few more 'clicks': 45 or 62
Striking & chiming – rack operated	45 Assemble front-work & operate going train to rotate minute wheel	Lifting piece does not lift } Lifting piece not reliable }	Repair minute wheel, pin or lifting piece: 46
		Lifting piece works every time: 46	
		Pawl does not release rack	More lift: 46
	46 Lift pawl clear of rack	Pawl clear, rack does not move	Check rack post & spring: 46
		Rack moves, correct warning: 48	
		Rack moves, but strike is immediate, no warning: 47	
	47 Check position of warning flag	Flag does not reach warning pin's path	Adjust relationship between flag and pawl: 48 or 57
		Flag too high	Ditto: 48 or 57
		Flag does not come free of warning pin	Adjust angle of flag, remove pits: 48 or 57
		Flag bounces off pin	Adjust flag angle: 48 or 57
	48 Check gathering pallet	Pallet holds the rack after lift	Under-side of pallet rough or pitted, stone it: 49

Clock type, train or part	Tests	Results	Correction
		Continuous striking as pallet bounces off, or passes rack stopping post	Pawl to be located better in rack teeth, or pallet tail too short: 49
		Continuous striking as pallet fails to gather	Pallet tooth too short or pawl not positioned correctly: 49
		Pallet gathers: 52	
	49 Check clearance between pallet tail and pawl	Pawl jams on pallet, no strike	Correct pallet: 50
		Pawl operates correctly: 52	
	50 Test strength of rack spring	Too weak, rack does not move	Adjust spring: 51
		Too strong, pawl cannot hold rack in position. Continuous striking	Ditto: 51
	51 Operate gathering pallet	Pallet gathers and stops on rack stopping post: 52	
	52 Fit snail and set for 1 o'clock	Operation of front-work correct up to gathering pallet, train halts just after warning	Adjust hammer loading: 53
		Rack does not move one complete tooth	Adjust position of rack tail: 52
		Strike correct: 53	
	53 Set snail for 11 o'clock	Train strikes 10 or 12	Correct position of pin on rack tail: 54
		Strike correct: 54	
	54 Set snail for 12 o'clock	Train strikes 11	Keep tail clear of hour pipe: 55
		Strike correct: 55	
	55 Operate snail at each hour	Strike variable	Adjust the clearance between rack tail & face of snail: 62
		All correct: 62	
Striking & chiming – locking plate (count plate or count wheel)	56 Lift & warning tests as for rack strike	Follow rack-strike route to 47: 57	
	57 Does locking hook drop into locking space?	Locking space is under hook and train is stopped: 58	
		Locking space is not under hook when train stops	Move hoop wheel in relation to rest of train, or bend locking hook (British, American and German clocks): 58 Move stop wheel in relation to rest of train: 58 (French clocks)
	58 Does locking hook touch either side of locking space?	Yes, train warns and then jams	Bend hook (Br, Am, or Ger)
		No: 59	Move stop wheel (Fr): 59
	59 Is hook falling into space & resting hard on bottom?	Yes, stopping piece will probably not operate reliably	Adjust relative angles of stopping piece and locking hook: 60
		No: 61	
	60 Is hook only just entering space?	Stopping piece will probably allow one strike only	Ditto: 61
	61 Inspect face of stopping piece	Piece may bounce out of the hoop wheel space	File face radial to hoop wheel with a chamfer at the bottom: 62
		Striking operated reliably: 62 or 65	
Hands – leaf-spring drive	62 Fit leaf spring and hands	Hands too loose	More bow in leaf spring: 63
	63 Check space between leaf & front plate	Clock stops	Correct leaf spring or fit a small extension collar to pivot: 64

Clock type, train or part	Tests	Results	Correction
	64 Check space between cannon pinion and bridge	Clock stops	Adjust thickness of washer between hands and their retaining pin: 67
		Hands & clock operate correctly: 67	
Hands – French-style drive	65 Fit pipe and hands	Hands too loose	Remove pipe and gently squeeze the sides in: 66
	66 Check space between cannon pinion and front plate	Cannon pinion rubbing on plate, clock stops Correct: 67	Adjust thickness of hand retaining washer: 67
	67 Fit dial after removing hands. Verify space between snail, hour or minute wheel	Interference between dial & front-work, clock stops	Adjust position of minute pipe, check posts & flatness of levers: 68
Dial	68 Refit hands	Clock will not go Clock goes for ½ hour or 1 hour: 55 Clock stops with date change Clock goes: 69	Check interference of hands: 69 Adjust date pin & wheel: 69
Case	69 Mount mov'm't in case & check for interference	Clock does not go Clock goes: 71	Adjustments to seat board or mountings: 70
	70 Check beat of crutch		
Pendulum	71 Inspect all parts before mounting; put in beat	Clock goes: 74 Does not go: 72	
	72 Does bob roll?	Clock unreliable	Suspension not held closely, adjust: 73 Suspension too tightly held, not plumb, adjust: 73 Crutch not square to path of pendulum, adjust: 73 Back-cock flexing, remove & correct: 73 Rod distorted badly, bob at an angle to path of pendulum. Remove & adjust: 73 Mounting of movm'n't insecure, adjust: 73
	73 Check lateral mov'm't between crutch and pendulum	Clock stops after several minutes Clock goes correctly: 74	Make crutch an easy, sliding fit with pendulum: 74
Finish	74 Fit hands and wind fully	Clock stops: 75	
	75 Check clearance between minute and hour hands and dial along their full length; test at all positions between 1 o'clock & 12 o'clock	Clock goes: 78 Clock stops: 76	
	76 Check space between hour pipe and back of minute hand	Clock goes: 77 Clock stops	Make clearance between hour pipe and minute: 77
	77 Check space between hands' arbor & clock glass	Clock goes: 78	

NOTE: It is just possible that, having reached this stage, your clock still fails to keep going. This intimates that either dirt has found its way in, or a faulty part has succeeded in passing its test and failed later in the sequence. The only thing that can be done (after looking for the obvious, ie not wound up, pendulum not entered on the crutch etc), is to carry out the preliminary tests again so as to establish the general area of the fault – case, display, pendulum, platform escapement or movement.

78 All done!

2
Cleaning Clock Movements

Before cleaning the movement it will be necessary to take it out of its case. Here are a few notes on doing this to various clocks.

Longcase Most hoods slide forward, but not all; so be careful as you pull the hood towards you. If it does not move easily, look first for a locking piece that is operated within the case and which locks the hood in position, and then try lifting the hood.

Next remove the pendulum. Remove the weights last and grip the cords as you take them off, so that there is no chance of the movement falling forward. It is a good idea first to put something soft on the floor that you can drop the weights onto.

Vienna and other wall clocks The method of fastening many wall clocks is to use 'keyhole' plates and screws. If the screws used were undersized, the case may be dislodged as the movement is withdrawn. Test the solidity of the case mounting first (Fig 29).

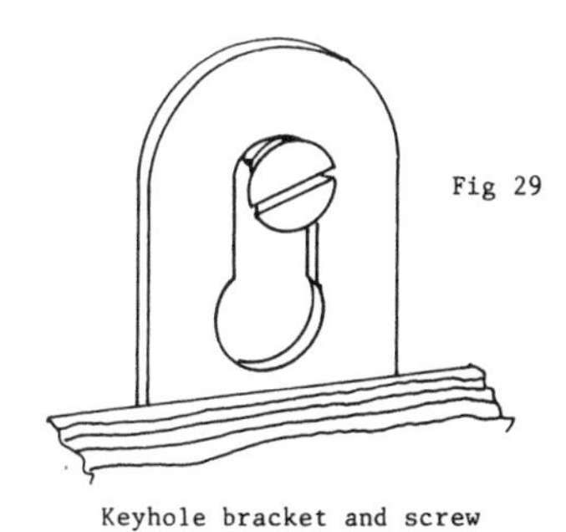

Fig 29

Keyhole bracket and screw

Carriage clocks Put a rubber band around the case before unscrewing the base and withdrawing the movement; this makes it easier to stop the frame and glasses collapsing in a pile on the bench. Do not overlook the fact that the door will fall out anyway.

Bracket clocks and mantel clocks Many bracket clocks are held in the case by two straps and two screws; the movements tend to be heavy and it is far safer to use odd pieces of wood to support the movement before loosening the screws, than to to try and hold it up by hand while this is happening. The same is true of many good-quality chiming mantel clocks, although it is frequently easy to lay these face down on a cloth whilst the screws are undone. Gongs and bells come out before the movement, and often before the pendulum.

FIRST STAGE

Old clocks, such as longcase, are particularly exposed to dust and cobwebs. They can be very dirty indeed, and if they are put into the cleaning fluid straight away will probably make it so filthy that it cannot be used again. As a matter of economy, therefore, it is a good idea to brush off the worst of the dirt, or even to put the whole movement in a large bowl and pour hot soapy water over it. Remove any very delicate parts such as automata first and, in the unlikely event of a clock with a platform escapement being in this condition, take the platform off. Lift the movement out of the bowl and put it to drain; make sure that nothing has fallen off into the 'soup' and then pour this away.

SECOND STAGE

This can either be carried out in an ultrasonic machine or by hand. The latter is the method used by those who do not clean many clocks, since the cost of the machine can only be justified by a reasonable turnover.

Dunking

This is the simple matter of taking a complete movement – assembled but without platform, pendulum or weights – and immersing it in cleaning fluid. Under certain circumstances this is a perfectly reasonable thing to do, but it must be remembered that if the movement is not dismantled the pivots cannot be re-polished, the pivot holes cannot be burnished, trapped grit will not be removed from springs and a number of faults may remain undiscovered. Cheap movements and completely open movements that show no indications of wear on the train are the only ones that ought to be cleaned in this way. Open movements include those going-barrel clocks where the end of the barrel has a large hole in it or the barrels can be removed without dismantling the rest of the movement (Figs 30, 31).

Cleaning fluids used for dunking must have a rust inhibitor in them and in all probability this will leave them slightly sticky. It is not a good method for movements that are not protected from dust. White spirit and paraffin (kerosene) with an addition of good lubricating oil in the proportions 1 part of oil: 30 parts of white spirit or paraffin, is a useful cleaning mixture.

Dismantling movements

Weight-driven movements present no dangers, but spring-driven ones must be tackled in a proper fashion or severe damage can be done to the movement and to the person dismantling it.

The movement should be run down as a first measure. The going train can be run down by simply removing the escapement, after locking the train with a thin piece of pivot wire through a crossing, and then letting it run whilst you rest a finger on the last arbor in the train and prevent excessive speeds being reached. Excessive speed will be noisy and self-evident. The other trains – striking and chiming – must be let down by carefully using a winding key to hold the winding arbor, slipping the click out of engagement with the ratchet and allowing the barrel to turn the key until you need to engage the click again and take a fresh grip on the key. Fusee barrels frequently need a spanner rather than a key, because of the size of the barrel-arbor square. Special tools for letting down are available and consist of a screwdriver-type handle with a fitting for the

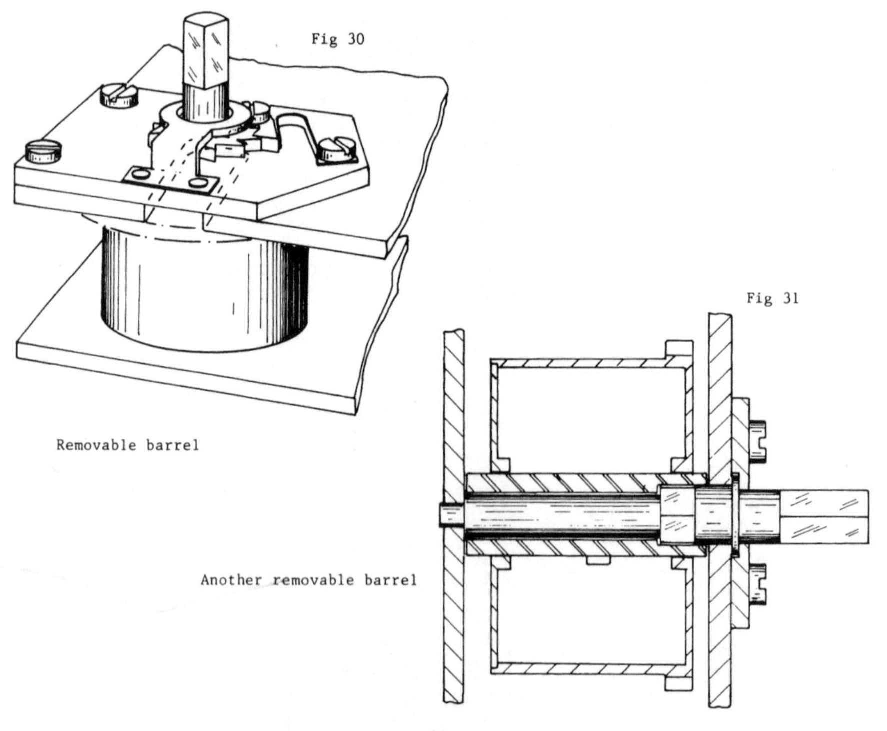

Fig 30

Removable barrel

Fig 31

Another removable barrel

square; I prefer to use a star-type key which gives a great deal of leverage over the barrel spring.

When the barrels are completely run down, remove the clicks or ratchet wheels to make the movement completely safe. If this is not done it is always possible that the spring has only unwound in part and that there is energy still in it caused by sticky coils. When the plates are taken apart this energy can suddenly be unlocked with broken pivots and other damage as a likely result.

If you are unfamiliar with the movement being dismantled, or if the parts are arranged in a complex fashion, always make rough sketches of the original assembly; this can save a great deal of time later on.

Cleaning by hand

Provide yourself with two plastic containers large enough to hold the clock plates as well as the other parts after totally dismantling them, and also a bucket of hot water. Put the clock parts in one container and fill the other with cleaning fluid. There are a number of proprietary brands available and you should experiment to see which suits you best. I prefer water-based cleaners – a mixture of water, soft soap and ammonia – for hand cleaning, and use Horolene which does not have the pickling effect that some others seem to have. Pickling ruins the surface of a good clock and removes sharp edges.

All the parts except the train wheels may be put into the fluid and left there for an hour or more. The great wheels may be put in too, but exclude the other smaller wheels as it is possible that the blanks from which these were made were hardened by hammering and this makes them prone to attack by ammonia. The form of attack is a corrosion that follows the grain boundaries, resulting in a crazed network of cracks; it only occurs in stressed brasses and hammering blanks results in stresses being left in the wheel. Typical failures due to this type of corrosion are crossings that break away from the rim, rims that break and teeth that fall out. A simple heat treatment which takes place at about the temperature of molten soft solder removes stresses, so that parts that have been heated since manufacture often do not suffer.

The practice of hammering blanks was not followed after sheet brass in half-hard condition came into common use, so that this type of damage is only a risk in antique clocks of great age. For mid nineteenth-century and later clocks it seems perfectly safe to clean all parts in ammonia cleaners.

If your skin is at all sensitive, or if you dislike fishing around in mucky cleaning fluid, wear rubber gloves. Help the cleaning process along by means of a little rubbing and then, if the parts seem free of verdigris, grease and the worst stains, take them out of the cleaner and put them immediately into the bucket of hot water. Any long delay between taking them out of the cleaner and immersing them in hot water will result in staining, which can be quite difficult to remove. Wash the parts in the water, then take them out to drain and dry them with a soft cloth. I am not keen on the normally recommended dry sawdust; it gets in the back of your throat and also it is something else that must be cleaned out of the holes in the plates.

The wheels can be cleaned mechanically. If you have a lathe or any other method of spinning them, use a piece of plastic to support emery paper and as the wheel spins, clean the surface by holding the emery paper flat against it (Fig 32). The support will ensure that the emery does not remove the crisp, sharp edges of the crossing-out. Reach into the shoulder of collet and wheel by folding a sharp edge into the emery paper and 'pointing' it into the joint. Finish with very fine emery paper or crocus paper.

Fig 32

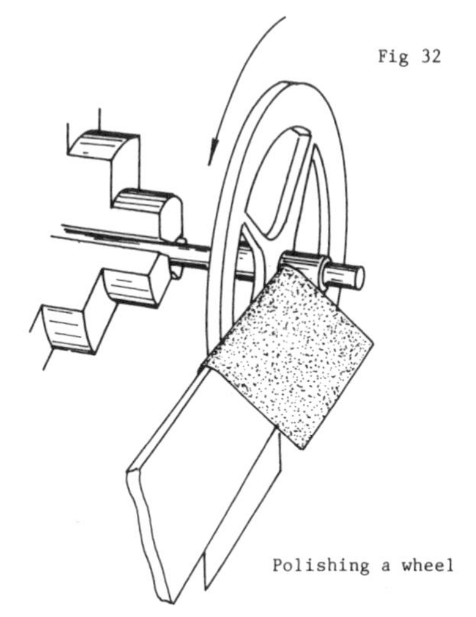

Polishing a wheel

Put all the parts back into the dry container. The cleaning fluid can be left to settle for an hour or so, and then all but the sludge at the bottom poured into a bottle for another day. Keep it covered – ammonia is a gas and vaporises readily. Topping up the fluid with commercial ammonia solution (35 per cent is the strongest that you can obtain) is a good way of extending the life of the cleaner.

Ultrasonic cleaning

The use of ultrasonics to clean clock movements can greatly reduce the time needed to clean and service low-priced antiques and modern clocks. The amount of time saved in cleaning a valuable antique – anything over about £250 – is much the same; but since more work on mechanical details will probably be involved, the proportion will be smaller. Again there are any number of cleaning fluids on the market. The one I use has ammonia in it, though not in a water-borne form; and neither I nor the people who recommended it have had any stress corrosion in more than five years. Nevertheless, I would not risk the wheels of a good London-made clock.

The usual method of cleaning is to use at least two fluids, one for the actual cleaning and one for rinsing. I keep a much-used batch of cleaner for the first attack on filthy clocks, a tin of cleaner that is new or at least in good condition, a tin of rinse for the first rinsing and a tin of final rinse that has a high rate of evaporation. Thus a very dirty clock goes through four fluids after the first crude hot-water treatment, and a clock in reasonable condition through only three. It is a matter of simple economics; unless you have a filtration system for the fluids, a dirty clock can make such a mess of a new cleaning fluid as to make it useless for normal work. Keep the fluids in tall bottles and allow the solids to sink before pouring off the cleaner material for the next cleaning job. The mud at the bottom (it should not be as extreme as that) can be thrown away. Since most of these fluids that do not contain water become jellylike when brought into contact with water, do not pour them into the normal domestic drain. Burn them or pour into an ash heap with no risk of drain-off.

Instructions will come with the ultrasonic tank and there is no point in reproducing them here; but a note of a few things that you can do, and some that you cannot, may prove useful. If the tank is large enough it will clean some assembled movements quite well; refer to the earlier comments on 'dunking'. Any weight-driven clock that does not require to be stripped can be cleaned entire.

Spring barrels must have the cap removed when cleaning – unless they have a large hole in the bottom – so that the fluids may be properly drained out and the spring lubricated. Even so, if there is any suspicion of grit the spring must be taken out and cleaned; an ultrasonic tank cannot remove grit from between touching coils.

Platform escapements should be cleaned in watch-cleaning fluid. If the fluid does not contain a lubricant, the platform will have to be stripped anyway; it might just as well therefore be stripped for cleaning. Oil may be fed into a cleaned barrel by means of a small syringe or an oiler. Penetrating synthetic grease is now available for jobs for this type. The grease, which is loaded with micronised PTFE, is supplied in a solvent that enables it to seep into the spaces between the coils. It then evaporates and leaves the spring evenly lubricated. Normal grease needs to be applied to the spring before installation which not only makes the spring difficult to handle but admits the possibility of picking up grit and dust. Penetrating grease avoids these problems.

Platform escapements must never be left on the movement for cleaning. They should be treated separately and rinsed in fluids that have not been used previously, so that there is no possibility of introducing tiny particles or a film of oil into what is virtually a watch movement.

THIRD STAGE

After washing the movement, whether by hand or ultrasonics, it must be inspected and finished by hand. A clock that was cleaned without dismantling does not leave much scope for hand work; it will be a matter of cleaning off any staining with a soft cloth, and using pegwood or a sharpened matchstick to clean out any odd crevices that still retain dirt. Inspection will largely be to ensure that the removal of grease has not revealed wear in pivots or pivot holes that was not suspected before. The correction of this type of wear is dealt with in Chapters 3 and 4, but here are a few points to help in recognising the various types.

A worn pivot will only show if it is so badly worn that it is waisted. Since the end of the pivot should protrude from the hole, the latter should be of the original diameter. If a magnifying-glass is used, it will be seen that the end of the pivot overlaps the edge of the pivot hole as the arbor is moved from side to side (Fig 33). A hole can only wear to an oval shape, in a clock; if it is not oval it has not worn. Ovality will be seen more clearly in an assembled clock by using a glass and moving the arbor from side to side, and then up and down – in other words testing the diameters at 90 degrees to each other. If there is more movement in one plane than in the other, the hole is oval.

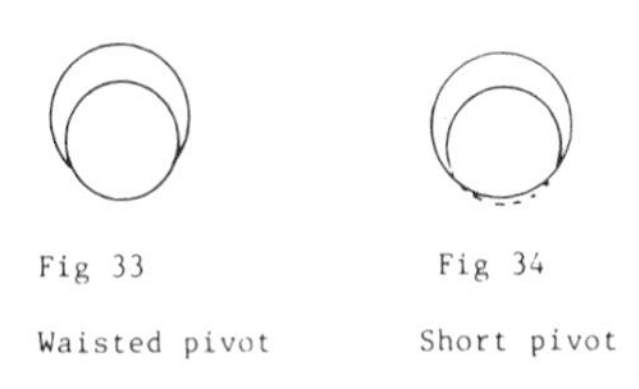

Fig 33 Waisted pivot

Fig 34 Short pivot

It sometimes happens that a pivot has been made too short, so that the end lies within the hole (Fig 34). I have never seen a short pivot that is worn; it always seems to be the hole that has suffered.

Inspecting spring barrels

Barrels that are removed from the movement should be inspected by lifting off the cap. If the underside of the latter is gritty with metal scraped from the barrel or from the cap itself, or with any of the substances that seem to creep into clocks, there is no possibility of cleaning the spring properly without taking it out. This does throw doubt on the quality of cleaning carried out when it is decided not to remove the barrels from a movement. Undoubtedly particles of metal or whatever will be clamped between the coils of the spring and no method of cleaning that does not open up the coils is going to do any good at all.

If there is no sign of grit, and if the old oil is merely thick – or non-existent – the spring can be washed out whilst remaining in its barrel, for a cheap repair. Put the barrel under a strong light and examine the anchorage of the spring to the case of the barrel; the sides of the spring at this point must be level with the groove in the anchor. If they are not, the actual part of the spring that is gripped by the anchor has 'lipped': it has been bent up and out of the line followed by the sides of the spring (Fig 25). That, in turn, means that there are small fractures around the anchor-hole of the spring. It will have to come out.

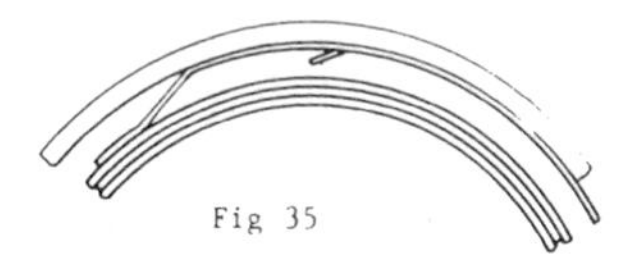

Fig 35 Spring with cranked end

Pre-war barrels in British and German mantel clocks often have the anchor pressed out of the metal of the barrel. Since this leaves rather more of the anchor sticking through the outermost coil and bearing on the next, the maker has often pressed a 'joggle' onto the very end of the spring to act as spacer between this end and the next coil (Fig 35). This makes the job of entering the spring easier and, more importantly, prevents the anchor depressing a very localised area of spring and producing a stress concentration.

Removing and inserting springs

To the best of my knowledge there is no completely safe spring winder. Even those with a clamp for the spring end, enabling the anchor of the barrel to be unhooked, have been known to let go and hurl the spring across the workshop. This is an area in which there is a lot of disagreement between clockmakers. Many assert that a spring will be distorted if it is put into, or taken out of, a barrel by hand. Others, including myself, claim that as long as the operation is done properly, there is no effective distortion. But I do not believe that this holds true for the much smaller springs of watches.

Distortion in a spring may be judged by holding it by the outer end and letting it hang vertically. If the centre coils 'belly out' to one side or the other by more than about 6.5mm (0.25in) on a spring from a French round movement (and pro rata on other springs), it is distorted. Having said that, it is true that if you cut the wire on a new spring and carry out the same test, you are likely to find the figure exceeded. So what is the effect of distortion?

If the spring does not naturally keep within its place when winding and unwinding, there is a possibility that the coils will bulge against the end cap of the barrel or the inside of the barrel end, and of course this produces scraping and increases the friction. In addition it is claimed that the spring will force the cap off the barrel. The first possibility is undoubtedly true; we have just looked at the problem of scrapings in the barrel. However, springs that have scrapings are often not distorted more than the amount stated above. Since it is anchored at both ends, distortion cannot account for the spring protruding further than the starting position of the innermost coil; in any case, if the anchor slot allowed so much movement it would only be part of the very innermost coil that could touch. I believe that the only reasons for the spring touching the ends of the barrel are dirt between one coil and another so that the coils are slightly tipped over to one side, or distortion of the side of the spring by dents or waviness which will also throw the spring over.

As to the second claim – that the cap can be forced off – there seems no prospect of belly producing sufficient force to do this. Dirt and dents are quite capable of generating a strong longitudinal force, and it seems more important to look for dirt and damaged edges than to worry about belly. Fraze is damage also, and a spring that has scraped because of dirt has very likely got fraze along one side or the other which should be removed. If the arbor is eccentric in the barrel cap or end, this will also create a longitudinal force.

SPRING WINDERS

There are two patterns of spring winders. One has a winding spindle in a frame and a bar over the top of the winder that carries an anchor for the spring; in the other the spring is wound into a cylinder through a longitudinal slot in the cylinder's side until only that part of the spring that anchors onto the inside of the barrel is left outside. This is then hooked onto the barrel and the spring can be withdrawn from the winding cylinder (Figs 36, 37). In both cases it is usual to wind the spring onto its own arbor, which is held in the winder spindle. Various models may combine these patterns, and at least one has a pair of narrow-bladed clamps that can be fixed to the outer coil of a spring so that it can be unhooked and the spring extracted; instructions are included with the tool. An important feature of all winders is a ratchet wheel to hold the spring in the wound condition.

HAND WINDING

The important word is 'winding' – the spring must be wound in or wound out of the barrel, it must not be pushed or pulled. To wind out, take a pair of smooth-jawed pliers and spread a duster-sized cloth over them. Grip the inside coil of the spring through the cloth with the pliers, and then take hold of the barrel with another cloth. Wind the pliers round to tighten the inner coil and gently draw outwards, taking up the movement that the winding allows. When the coils have been screwed out far enough to grasp by hand (through the cloth), apply the twisting to the extended and telescoped spring. Continue until the spring begins to come out of its telescoped state, the cloth will tangle into the coils of the spring and assist in letting it unwind slowly. When it is right out release the spring from the anchor by sliding it around the inside of the barrel.

Putting the spring into the barrel requires a little more exertion. Hook the spring end onto the anchor making sure that it is the right way round, then grasp the spring and bend it into its natural curve, easing it into the barrel. Work around the barrel, gradually bending the spring in and being careful when you pass the end not to slip into the space between the spring end and the barrel. As your fingers bend the spring, your thumb should be supplying the movement to coil it tightly in – the first two or three coils are the worst, after that the spring is not so insistent on escaping again. Use the cloth to hold the barrel, and perform the whole operation so that it hangs around the spring like a screen. If you do let go, the cloth will both catch the spring and protect you.

WINDING ON THE LATHE

Some springs require more muscle than most of us are endowed with, but if you have a lathe with a geared headstock or a back-gear (the old-fashioned slow-speed arrangement of a belt-driven lathe) it can be used to wind springs. Provide a support in the tailstock for one end of the arbor, making sure that the

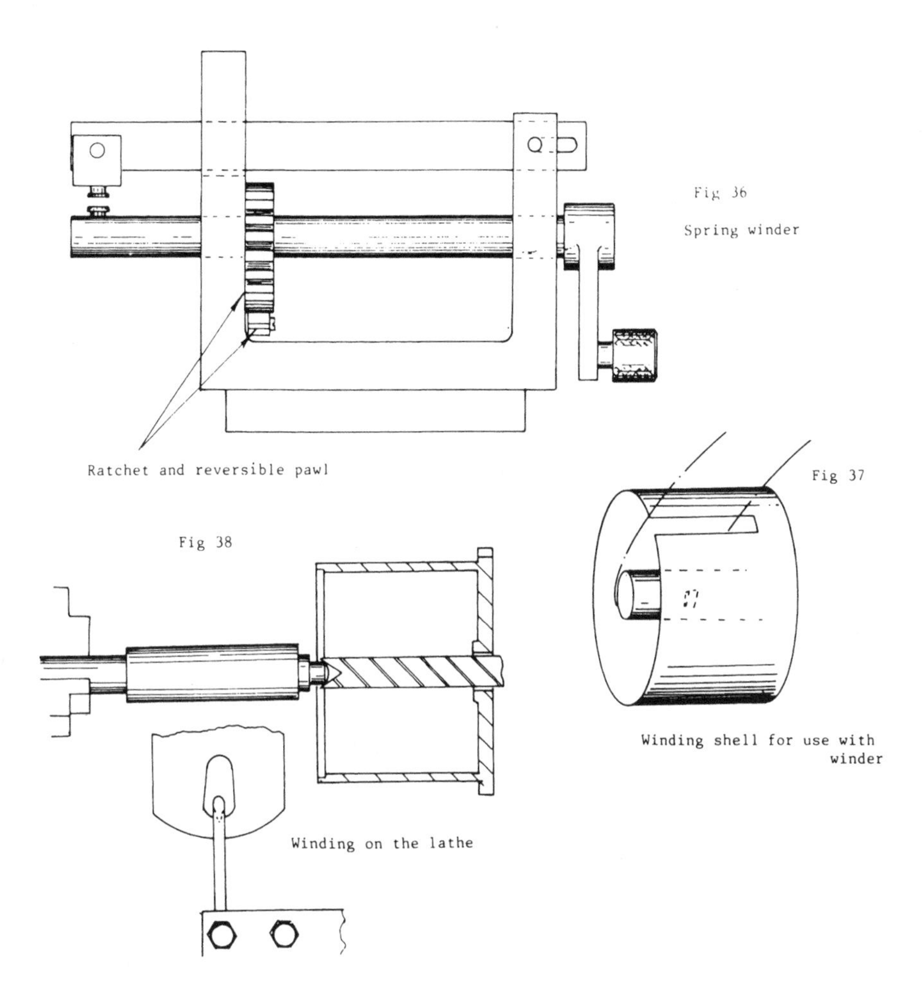

barrel can be pushed back far enough over the arbor or the support to fully expose the part of the arbor that the spring is mounted upon. Grip the other end of the arbor (avoid the pivot) in the chuck, and place the barrel on the arbor and/or the support. Make a hook out of silver steel, either fitted with a 'tee' handle for holding firmly, or made to clamp in the tool-post. Now all that is necessary is to hook the inner end of the spring onto the arbor, wind the chuck in the right direction by hand, holding the slack of the spring in your hand until you need to let the hook take the strain (Fig 38). When the spring is wound small enough, slip the barrel as far over it as it will go, and gently let the chuck wind back under spring power until the spring is tight in the barrel. Remove from the machine and tap the spring down in the barrel with a sheet of scrap brass and a hammer. The spring must enter the barrel far enough to avoid fouling the anchor when it is tapped down. This is the method I use for really big springs such as the big three-barrel ones in nineteenth-century director's clocks.

AMERICAN OPEN SPRINGS

These are usually either open ended or riveted and loop ended. The former is the easiest to replace – simply hook onto the arbor and wind in through the assembled mechanism. Loop ends are more awkward. New ones are generally supplied in a C-shaped clip which enables the repairer to simply slip the spring over arbor and anchor-post when the clock is being assembled (Fig 39);

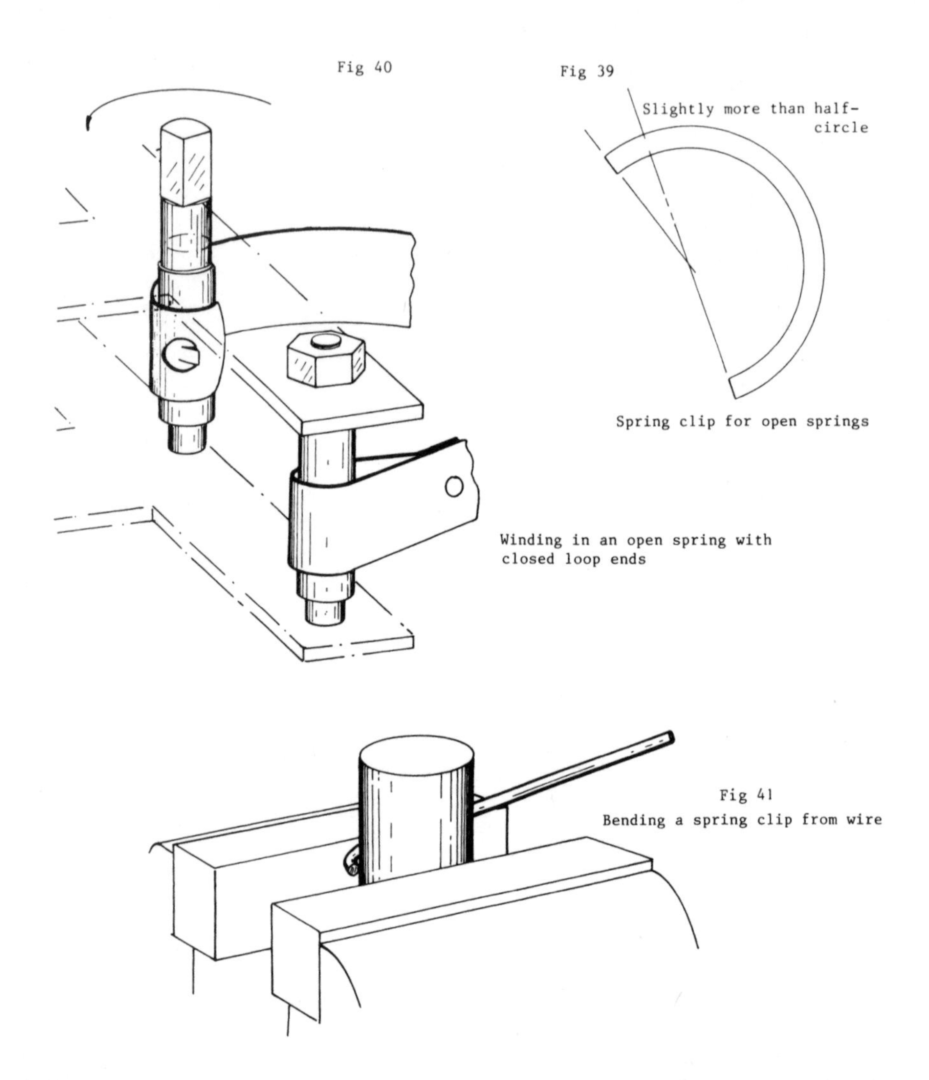

Fig 40

Winding in an open spring with closed loop ends

Fig 39

Spring clip for open springs

Fig 41

Bending a spring clip from wire

the clip is released, when assembly is complete, by winding the clock up completely. My own method for an uncoiled spring is to straighten the spring out in my hands until I can lead the inner coils onto the arbor and pop the riveted loop over its post, leaving the rest of the spring outside the movement (Fig 40). The coils will twist over each other, but will come to no harm. The clock is assembled and then wound up while I untwist the coils and ease them into position.

It is not often that one has to remove an undamaged open spring, but when you do, use a 'C' clip which can be mounted very easily whilst the spring is in the movement. The clips can be bought if you do not have a suitable one from an earlier purchase of a new spring, or they can be made from piano wire obtainable from a model shop, or high tensile fencing wire from a garden centre (Fig 39). The diameter needs to be about 2.5mm (0.1in) and, since the wire is quite difficult to bend, the vice should be solidly mounted on a good bench. The radius of the inside of the clip should be small enough to allow the spring and clip to be slid into place, but large enough for the spring to have a little more 'wind' left in it so that the clip can be removed afterwards. Find a piece of steel bar of the diameter that the clip is to have on its inside, and grip the bar in the vice with the end of the wire trapped between it and one vice jaw; the rest of the wire can now be bent around the bar until about three-quarters of a full circle is obtained (Fig 41). Take the bar out of the vice and cut the clip from the curved wire so that a part-circle of just over 180 degrees results – anything between about 200 degrees and 240 degrees will do.

Repairs to springs, barrels and gut barrels can be found in Chapter 13.

FINAL CLEANING FOR ASSEMBLY

Most plates will be quite clean enough after treatment in cleaning fluids, but occasionally there will be damage to the plate from bad handling in the past. The only way in which scars and fraze can be removed is to carefully file them away; the same is true of bushes that protrude from the plate surface. However, before putting a file to the surface, you must decide whether any important information such as repairers' marks and dates, makers' marks, tool marks and of course, engraving, will be destroyed by the small amount of metal that will be removed. Fortunately the removal of metal is nearly always a localised affair, but look carefully first. All file strokes and polishing must follow the same direction – preferably top to bottom – otherwise teeth and abrasion marks will show after polishing.

After filing, fine emery papers should be used, backed up with something firm so that the edges of the holes are not rounded over. Start with a grit size of about 150, proceeding to 300 and 600 before finishing off with crocus paper and Duraglit – a polish that does not leave powder behind. While using the emery paper you may find hollows that cannot be touched with the backed-up paper. Clean these hollows with the paper held in fingers alone, but do not wander over any pivot holes if you can avoid it, and keep the same direction. Do not be too afraid of removing historical marks, it will not happen suddenly; it takes a great deal of hard work to remove even 0·01mm (0.0005in) and the scratch marks of repairers will be much deeper than that.

Pegging out

All the holes (and any odd corners) ought to be pegged out before burnishing to avoid the risk of the polishing tool pressing grit into the pivot-hole wall, where it can carve into the pivot surface when assembled. This is a simple matter of using a pointed piece of wood to push grit and dirt out of the holes. Materials suppliers sell bundles of pointed sticks (pegwood) for this purpose, but I prefer to sharpen a matchstick and use that for all but the smallest holes (about 0.4mm or 0.015in). Matchsticks are soft enough to allow the grit to become embedded in them and they are impregnated with paraffin wax which leaves a lubricated surface behind. After pegging, use a polishing tool to burnish the holes and then peg out again. Finally, lightly chamfer the inside end of the hole where it breaks into the inner surface of the clock plate. It only needs to be a hair's breadth, and is intended to ensure that the hole and the corner of the pivot shoulder cannot bind together. Wear cotton gloves during final polishing, or handle with a duster.

Lacquering

It is sometimes desirable for the plate to retain its polish for several years even though exposed to the atmosphere. Carriage clocks do not often have their plates lacquered, but they are well protected. Some very good lacquers are available from materials suppliers for protecting polished brass, but they must be removed from the pivot holes after application to the plate. Lacquering wheels is even more difficult and in the best clocks this problem of protection is solved by gilding the brass.

Lacquering is an art, it needs practice and a good lacquer that suits your circumstances. To be perfect a lacquer should cover readily, flow to give a smooth thin surface, and dry quickly before any dust can fall on it. Unless you can take practical instruction from someone who does it regularly, the best alternative is to use a good furniture wax, containing beeswax. I find that this will give a good finish for years, with the added advantage that when used to cover silvered dials it does not become scratched by hands' adjusting or by cleaning the clock (as lacquer does), but is spread evenly over the surface when a duster is used. It can be renewed easily once or twice a year in the course of normal furniture polishing.

3
Bushing Clock Plates

The reasons for bushing the pivot holes in clock plates are as follows:

1 The pivots have been re-cut and are now too small for the holes.
2 The holes have worn oval.
3 The holes have been 'punched up' by some previous repairer.

If the pivots have not had anything done to them other than polishing, and the holes are truly round, then a clock that was running without giving trouble prior to stripping down does not need bushing. Even if the holes seem sloppy, this is almost certainly the amount of freedom that was built into the clock in the first place. Holes only wear in one direction in the normal clock and, consequently, as we have seen, the evidence of a worn hole is ovality; if the hole is still round it has not worn to any degree that falls outside clockmaking tolerances. Assuming that the plate needs bushing, the old holes will need to be drilled out using a drill that is sufficiently large to give a stout wall thickness to the bush – usually two to three times the pivot diameter.

Use a clear area of the bench, and cover this with a small sheet of paper (A4 or foolscap); lay the plate flat on this, inside uppermost. Using a steel rule and a scriber draw crossing centre-lines over the unworn part of the hole or, if it is for reasons (1) or (3), simply over the existing hole (Fig 42a). Since a hole that has worn is oval, the pivot has been moving away from one side of the hole ever since the wear began, and at least 180 degrees of the original hole circumference remains. This is what the centre-lines are set up on, it is a matter of judgement and must be done carefully. If after drawing these lines it is apparent that one, or both, the lines are not truly on centre, draw them again on the other side of true centre so that centre is exactly midway between both lines (Fig 42b). This is far better than attempting to correct the marking out and then having doubts as to which of two lines is the right one. By this method you always aim between any double lines.

The holes are now ready for drilling out (Fig 42c), and by far the simplest method involves a drill stand, or drill press. Select a centre (Slocombe) drill the same size or smaller than the proposed bush diameter, and fit it in the drill chuck. Lightly clamp the plate to the drilling table and 'sight' the tip of the drill to fall on both the scribed centre-lines, viewing from points A and B (Fig 43). Tap the plate from side to side until you are satisfied that the cross-lines are immediately below the centre of the drill, then carefully tighten the clamping. Check that nothing has moved.

Fig 42a

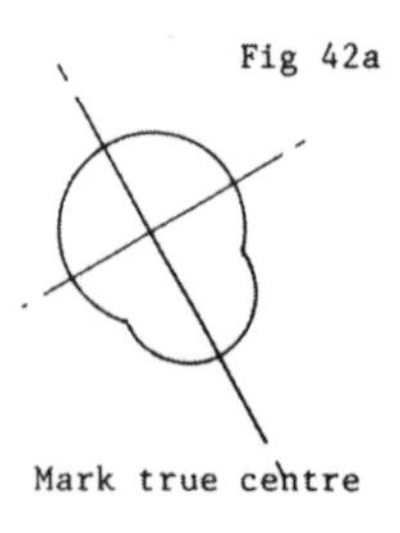

Mark true centre

Fig 42b

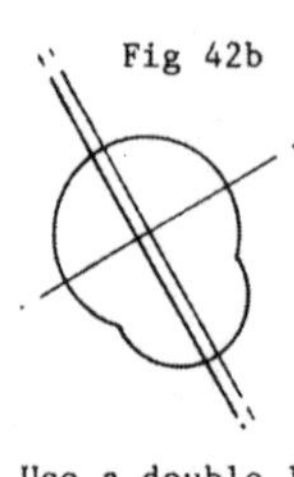

Use a double line when first is in error

Fig 42c

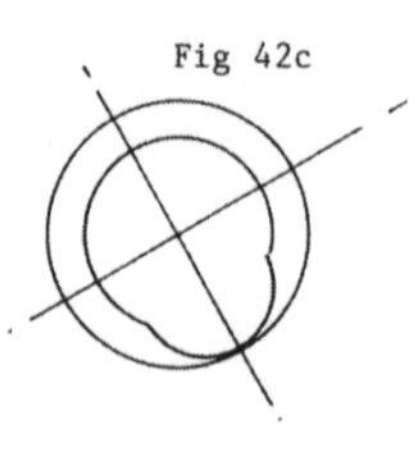

Bush hole true on marked centres

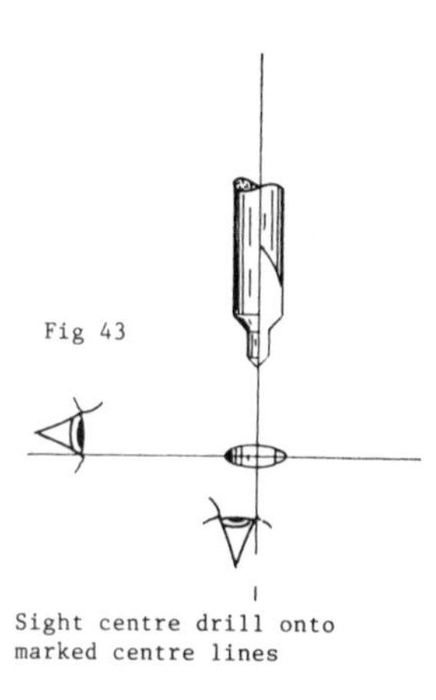

Sight centre drill onto marked centre lines

Set the drill in motion and gently take it through the plate. If the hole was oval, the stiff centre-drill will bore out the metal to give a hole concentric to your crossed centre-lines. It may be that the amount of wear was more than could be removed by the drill to leave a clean round hole; in this case, ignoring any previous decision as to the size of bush to be fitted, go to the next size of Slocombe drill to gain a completely round hole. A drill of this type does not have any circumferential clearance and needs oiling when passed right through the metal; treat it carefully and do not let it seize up in the hole.

Having obtained a true hole, open it up to the size that you had decided upon, or, if it is already to size, remove the plate from the machine. If no drilling machine with a firm table is available, the hole must be trued up first of all with a round file, using the marking as a guide. When it is true to the centre-lines, open up to size with a hand drill, or broach.

The following slight modification to the marking out makes trueing by file easier. Fill the worn hole with a small lump of lead or solder and hammer it level with the plate. Carry the centre-lines across the lead making a cross. Now put one leg of a pair of dividers on the intersection and draw a circle slightly larger than the hole that you intend to make. This circle will demonstrate conclusively the accuracy of the finished hole (Figs 42c, 44).

The plate now has a hole in it that is big enough to accept a bush and is either parallel, or slightly tapered. Put a piece of brass rod in the lathe, drill it to slightly smaller than the required pivot hole, and turn it until – with

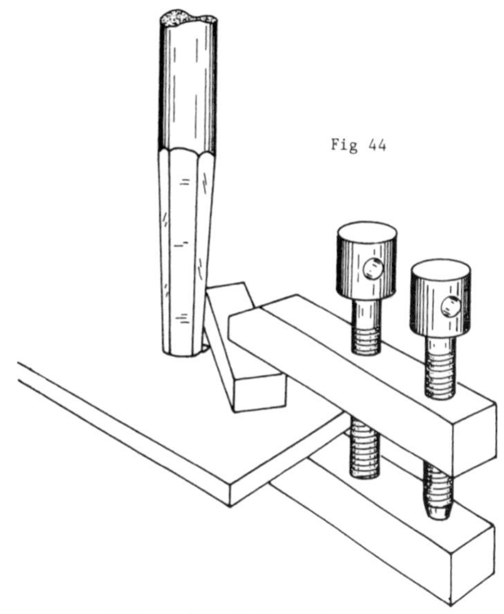

Using a broach to produce a bush hole true to centre

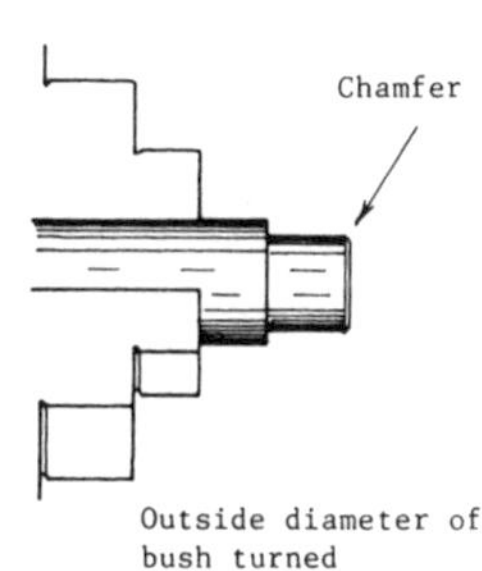

Outside diameter of bush turned

the sharp corner wiped off with a fine file – the bush just jams into the drilled hole in the plate. It should support the plate without entering the hole completely (Fig 45), making an interference fit of about 0.05mm (0.002in). Put a parting tool in the tool-post and, using a vernier calliper, a depth micrometer or the calibrations on the saddle traverse, part the bush off the bar to a length of about 0.25mm (0.010in) more than the thickness of the plate at the point of bushing (Fig 46); old plates are often not parallel.

A smooth-jawed carpenter's vice is the easiest way of pressing the bush home. Enter the bush so that it is fairly square to the plate and then put plate and bush into the jaws of

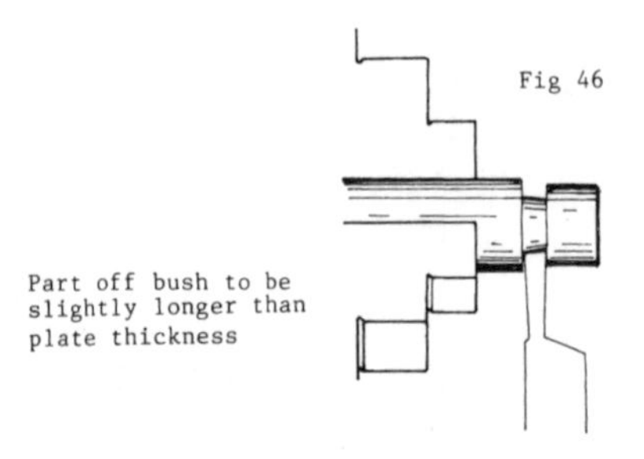

Fig 46

the vice and squeeze the bush into the plate (Fig 47). Make sure that the vice jaws do not scratch the plate by placing a soft metal guard on them or putting stiff card between the jaws and the plate. A light tap with a planishing hammer will complete the job so that the bush is just a little proud of each side of the plate.

If it is not possible to squeeze the bush home with a vice it should be tapped in with a light hammer and a smooth-ended punch, preferably a guided punch similar to that in Fig 48, and made as follows. Take a piece of brass large enough to drill a hole into and somewhat larger than the bush, and face the end in the lathe so that it is square to the hole. Now turn or select a short length of silver steel so that it will just slide easily through the brass and protrude about 12.5mm (0.5in). Make this into a flat-ended punch of the form shown in Fig 48, then harden and temper to a dark straw. If the bored-out metal is placed over the bush that is to be inserted into the clock plate, and is held flat on the plate while the punch is placed inside it and then struck with a hammer, the new brass bush will enter the plate squarely without suffering any bruising of the edges from careless hammer blows.

Place a piece of cartridge paper between the end of a single-cut flat file and the plate

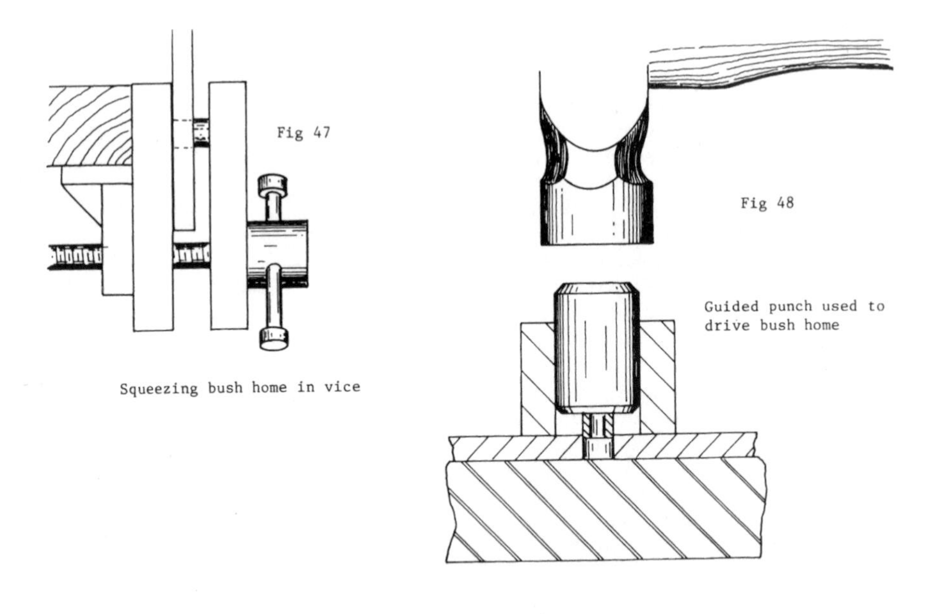
Fig 47

Squeezing bush home in vice

Fig 48

Guided punch used to drive bush home

Fig 49

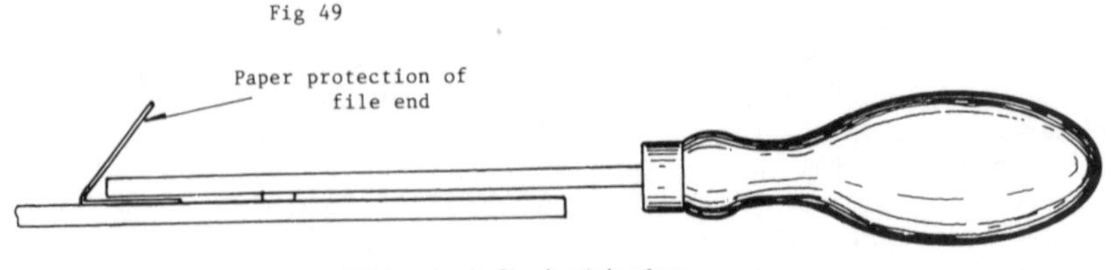

Filing bush flush with plate

surface, and carefully file down the face of the bush until it is no longer proud; repeat on the other side and finish with a block-supported piece of emery paper (Fig 49). All strokes must be made along the length of the plate wherever possible, so that no scratches will appear across it. The amount that was left on the bush length for subsequent removal was kept small so that the file would be kept nearly parallel to the plate when levelling off; this avoids the danger of unwarily 'pitching' into the surface and scoring it heavily. It also involves less work.

The whole operation takes less than ten minutes from marking out to polishing, and all that remains to be done is to size the pivot hole to suit the pivot.

The setting-up of clock plates for drilling can be alternatively done by slipping a pointed hard-steel rod into the drill chuck and using this to register the remaining original periphery of the worn pivot hole. The cone of the point used should be relatively shallow so that it does not obscure your view of the hole, an included angle of about 45 degrees works very well.

Mount the plate over a piece of wood with a hole beneath the pivot hole to accommodate the point of the locating rod, leave the plate unclamped but with the clamps in position. Lower the pointed rod into the pivot hole until it is bearing on the original periphery and use your hand to push the unworn part of the hole hard against the cone. When the cone is touching all round the unworn part of the hole, clamp the plates down, check that nothing has moved, remove the locating point and drill right through the plate with the centre drill as before.

Although this method is quick, the locating cone does tend to obscure the hole and anyone unfamiliar with setting a pillar drill will find themselves wishing for a third hand. If you have frequent bushing operations to carry out it is well worth using a simple toggle clamp that can be fitted to the drilling table and foot-operated by a cord and stirrup.

Old clockmaking practice was to use a tapered hole and a tapered bush. Mass-produced 'bouchons' are still made tapered, which means that a part of the 'squeezing' force is always acting along the axis of the bush and tending to push it out. The fitting of a bouchon must be done very carefully or it will pop out and, since the wide diameter is on the inside of the plates, will slide inwards against the arbor shoulder and very probably stop the clock without there being any obvious evidence as to cause.

Bushing wire is available, this being brass tube in short lengths having inside and outside diameters convenient for bushing and, above all, parallel. When no lathe is available, a hole is produced in the same fashion as described above, but sized by broach to suit the bushing wire. Unless the broach goes in to its full diameter the hole will be tapered but, so long as the bush is entered into the wide end, a slightly tapered hole will hold a parallel bush as firmly as a straight one.

The new pivot hole must now be broached out to suit the pivot, so select a broach that will cover the pivot size. Broaches are tapered and each size ranges over a series of holes; however, the large end of the cutting surface is parallel and many clockmaking books extol the use of this surface alone for finishing the hole. The problem is that although the idea is good, one would still need hundreds of broaches to satisfy all one's needs in clockmaking. Common practice is to use the broach's tapered surface to open up to the desired size, and to cut in from both surfaces of the plate so that a cross-section of the hole shows a very slight 'diabolo' form, wide at each end and narrow in the middle (Fig 50). Holes should be square to the plate, but in many old clocks movement has taken place between the set of holes on one plate and those on the other, so that it is worthwhile pinning the plates together in their nor-

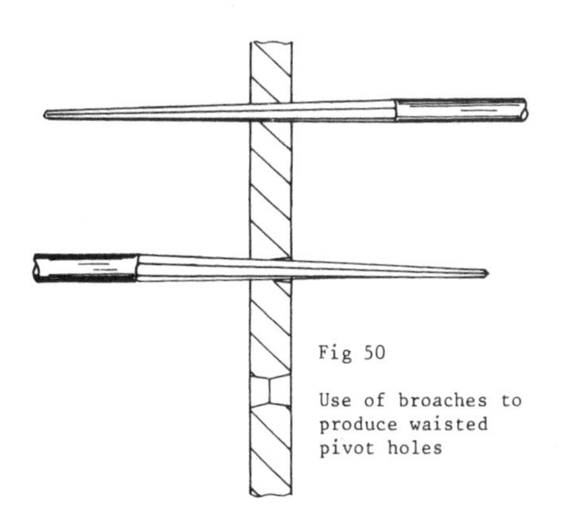
Fig 50

Use of broaches to produce waisted pivot holes

mal position, putting the broach in the hole to be opened and 'sighting' (Figs 51, 52). In this way you can ensure that the new hole is in line with the other. This is more important than getting the hole square to the plate, unless a complete restoration is being undertaken and all plate errors are to be corrected.

The amount of clearance between hole and pivot is extremely important, and a very common failing amongst inexperienced clock repairers is that of making the 'fit' of the pivots too tight. When a clock has been assembled, all the arbors should drop from one side to the other as the movement is turned over – it should rattle. The exception is a regulator of rigid and heavy manufacture; that would not be expected to rattle, but the arbors must still fall from one plate to the other with no hint of catching.

No more specific rules can really be applied. It is stated in various books that clearance should be sufficient for an arbor, when set upright in one plate, to lean 5 degrees either side of the vertical; and at least one American source specifies 1 degree. But since clearance is a function of pivot diameter it will be the same for any given diameter in plates of different thickness; Fig 53, however, shows that the angle of lean will be affected by the thickness. Plates move when the driving force of weight or spring is applied – only slightly, but this is why so much 'play' is desirable. In practice, clearance is often between 5 per cent and 10 per cent of pivot diameter. It is worthwhile measuring the difference between hole and pivot in old clocks that retain round holes and unscarred pivots. As previously stated, these holes are unworn, and the dimensions used have proved to be good since the last time the clock was serviced. An unworn bearing is a better argument for a particular clearance than any scheme of geometry.

After sizing the hole, use a cutter to produce the oil-sink, which ideally should be the same diameter as the bush. If it is a very large bush (used to remove traces of previous punching around a pivot hole) this cannot be achieved and the bushing will show faintly. Polish the hole, and remove the fraze from making the oil-sink with a cutting broach. Cutters for making oil sinks are dee-bits with hemispherical cutting faces; dee-bits and broaches are described in Chapter 13.

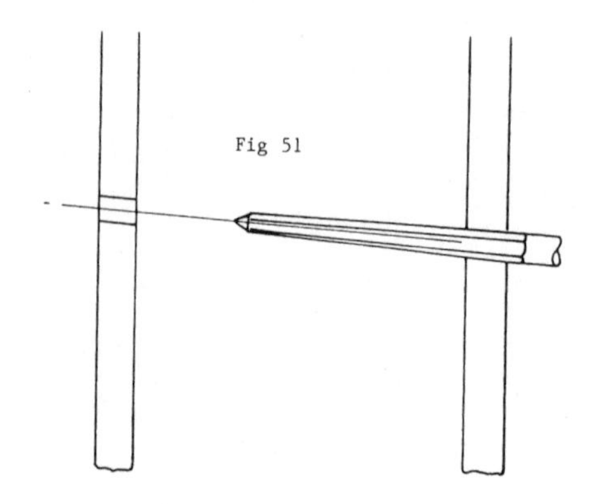

Fig 51

Exaggerated view of broaching pivot holes in-line

Fig 52

Broach end viewed through further pivot hole

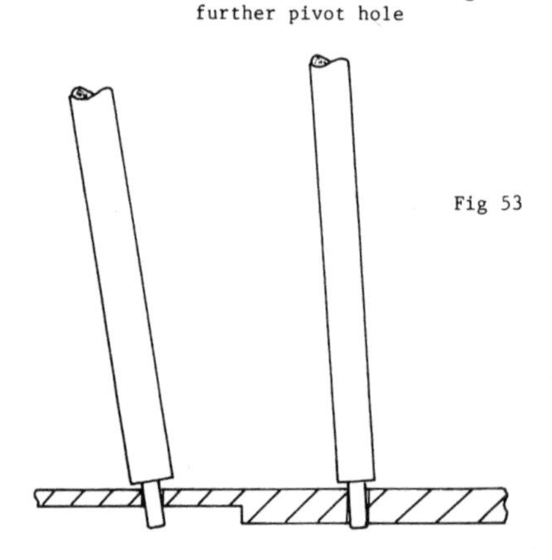

Fig 53

Effect of plate thickness on angle of pivot freedom

Good bushing technique can restore a clock movement to its original state, 'punching' is an abomination. Bushing has been used in new clock plates, usually to thicken the bearings in cheap, mass-produced clocks. It is very effective, indeed there is much to be said for the practice. Metal of good bearing characteristics can be used and, when at last the hole is worn out, the repairer has no need to mark out his centres again but simply presses out the old bush and installs a new one. However, most purchasers of a new clock would view such plates with great suspicion.

One often comes across older longcase clocks whose plates have raised, or upset, metal around the periphery of the great-wheel pivot holes. This served to increase the bearing area in relatively thin plates and to refine the grain in cast ones. Usually the grain in a brass casting is large, but when beaten or otherwise worked upon it becomes smaller and of a totally different form. In the alloys that are used in clockmaking, a small grain is tougher and of better bearing quality. The raised metal was achieved by hammering a tapered drift into a hole that was smaller than the desired finished diameter, and by supporting the plate over a hole that was about 6mm (0.25in) larger. The drift was placed in the hole from the inner side of the clock plate, and final sizing was by means of an orthodox broach. With the advent of half-hard and hard rolled-brass plates this technique became unnecessary, but was often followed by country clockmakers as a matter of established custom.

BUSHING SYSTEMS

No mention has been made of the various systems of reaming out plates on jigging fixtures and inserting manufactured bushes, essentially because I do not consider them faster than the use of a lathe and drilling machine. However, this is a personal view, and may even be influenced by an inborn parsimony for I dislike buying specialist machines that take up space and can only be used for one purpose. Many repairers swear by such methods.

Bergeon and KWM

Both systems have a table that allows the rapid clamping of a dismantled clock plate and subsequent sliding into true location under the reaming tool-holder. Sliding is in one plane only, so that location depends on positioning the plate correctly before clamping. Once located all tools are held accurately over the established centre during the reaming operations and the bushes too can be pressed home whilst the plate is clamped. This ensures that bushes enter squarely (Fig 54). All reaming operations are by hand, and the assembly pressure on the bush is approximately the same as can be applied by the use of a drill press (pillar drill).

A large range of bushes is supplied. In the

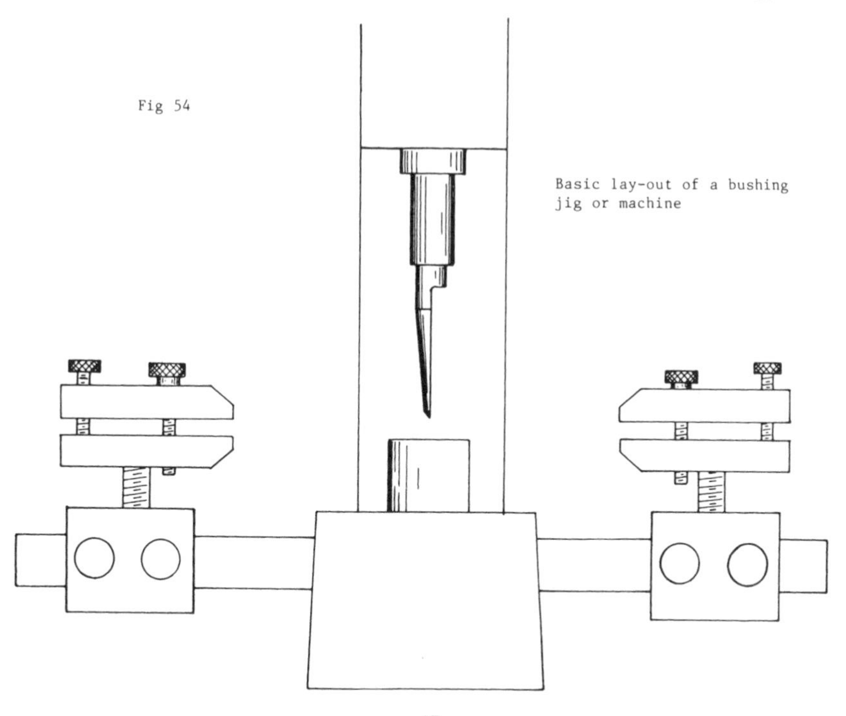
Fig 54

Basic lay-out of a bushing jig or machine

case of the Bergeon the number of different outside diameters is fairly restricted, amounting to nine only; but the outside diameters in the KWM system add up to more than twenty. The latter also caters for smaller pivot sizes than the Bergeon, but even within the same range of bores has many more outside diameters. Whether you view this as an advantage or not will depend on your feelings about stock, and how often you find that parts from one box manage to get into another containing a different size.

Non-dismantling systems

There are one or two methods of bushing clock plates without dismantling the movement – the Ayres for instance. It is inexpensive and consists of a guided support that clamps onto the clock plate. A hollow mill is used in this guide and locates over the pivot (a separate locating rod is provided), the mill is turned by hand and bores out the pivot hole around the pivot. Other tools are inserted in the guide to tap the hole and then insert a threaded bush.

Obvious advantages can be claimed for such a system. Apart from time saved in not dismantling the clock there is the matter of not disturbing parts that have worn together for decades. Nevertheless I do not like it, for the pivots are left unpolished and the bushes are anything but invisible. No matter how cheap a clock is, it seems unfair to treat it quite so cavalierly. Incidentally if several holes have to be bushed, I doubt that any time would be saved by not dismantling.

Cleaning a clock bushed in this way must incorporate a means of removing all traces of swarf produced in reaming and tapping – the most satisfactory method would be ultrasonic cleaning. Movements with spring barrels that cannot be removed without dismantling the rest of the train are not very suitable for this bushing method, since there is the problem of draining the barrels properly after cleaning and rinsing.

4
Pivoting and Mounting

The complement to the work described in Chapter 3 is repairing and replacing the pivots. If the original pivot is broken, or worn so badly that it has a visible waist, it is fairly clear that a new pivot is required (Fig 55). On the other hand, some faults are not at all obvious and can only be identified with the aid of a good magnifying-glass.

Fig 55 Worn pivots and badly made pivots

A very large proportion of the friction that is developed in the movement of a clock depends on the design and finish of the pivots and pivot shoulders. Probably up to 60 per cent of the total frictional loss in a longcase clock is attributable to friction at the shoulders alone. Tooth mesh, pivot diameters and the escapement are the other major components.

The pivots must have a polished surface at both the shoulder and the diameter, and (with the exception of early lantern clocks and birdcage movements) the diameter of the pivot must be parallel. The exceptions to the rule have tapered pivots – it appears to have been a change in practice – and since this is the original design, it should be retained. Clocks of this type must have the freedom (shake) of the pivot checked when the shoulder is lying against the inside surface of the movement plate, so that there is no doubt that the pivot will not bind if it moves over and engages its large diameter with the bearing of the pivot hole.

Normal pivots are truly cylindrical and not tapered or barrelled, because either of these characteristics can result in the pivot binding in the hole unexpectedly – unexpected because the fault will only occur under certain conditions which may never happen at the time of testing. If the pivots are not parallel, but strongly tapered, the arbor centre distances will vary according to what part of the pivot is bearing in the hole and the meshing of the gears will alter. This variability of mesh may not actually stop the clock, but it will certainly have an effect on timekeeping.

RE-SURFACING A PIVOT

Pivots are re-surfaced while still in the hard condition. It is not necessary to soften them first because the pivot file is quite capable of removing the small amount of metal needed to expose a clean surface once more. There are two or three different designs of pivot files offered by suppliers; Fig 56 illustrates the most common one.

Before starting to work, use a micrometer to determine whether the pivot is tapered or not. The faces of the micrometer anvils, being parallel to each other, will show up any variation from a true cylinder when the pivot is lightly pinched between them and held up to the light. For example, if a 1.25mm (0.049in) diameter pivot, 3.5mm (0.138in) in

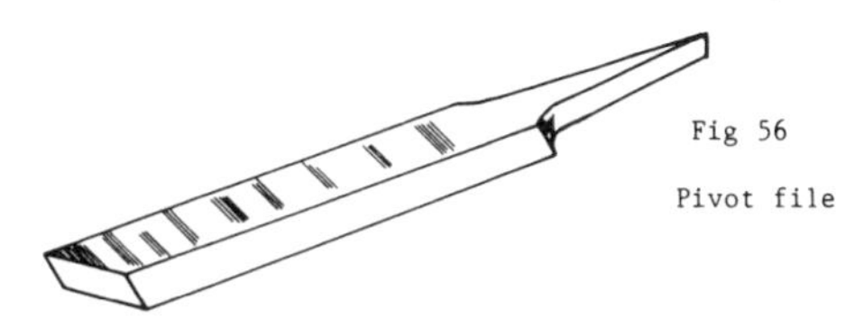

Fig 56 Pivot file

length, is tested in this way, light will show clearly between the work and the anvils if a taper of only 0.05mm (0.002in) is present. All manner of pivot deformation can be discovered in this manner.

Re-surfacing can be done with a file and a simple support in the vice. The support is a piece of brass strip or sheet, thicker than the diameter of the pivot, and with a saw cut at the top that runs the length of the face and locates the pivot as shown in Fig 57. The arbor can either be held directly between the fingers, or it can be gripped in a pinchuck; whichever method is used, you must use one hand to rotate the arbor so that the pivot is turned smoothly under the file held in the other. It is not easy to do without filing facets onto the pivot, and it requires a lot of practice. Usually this method is only needed for arbors that cannot be swung in the lathe because of the attachment of the crutch – in other words, the pallet arbor.

Most repairers will have either a lathe, a pistol drill or some other machine that can hold and spin arbors. If the pivot to be filed allows the arbor to be supported in a chuck as shown in Fig 58, the procedure is straightforward. Do not try it with pivots that are smaller than about 0.5mm (0.02in) diameter though, or you will very probably break them off.

Choose a pivot file that will bite into the corner of shoulder and pivot when it is held underneath the pivot. These files do not appear to have any rake on the teeth, and cut as well on the backstroke as they do on the forward one. Consequently, if the file is used on the underside of the pivot it will still cut well, and you will be able to see what part of the pivot is actually being cut because the file wears a bright ring on it (Fig 59). You should be able to cut right into the corner of pivot and shoulder and obtain a parallel pivot with very little trouble. Many pivot files are provided with raked sides so that only the edge touches the shoulder; others will need a little grinding along the side to achieve this clearance. Lathe speed should not be too great for filing this hard material; 500 to 750rpm will do in most cases, but if the metal is very hard and 'squeals', drop the speed. If the pivot proves too hard for the file, even at lower speeds, it has not been tempered; heat the arbor gently until the pivot turns blue.

A large proportion of the pivots in any clock will need to be supported for filing. This can be done using a Jacot tool (a fairly expensive piece of equipment), or by making the simple brass support shown in Fig 60a. Take a piece of brass rod and drill it to just a little larger diameter than the pivot, so that the pivot does not bind in the hole. Turn the outside of the rod so that it is small enough to hold in the tailstock drilling chuck, or whatever arrangements you have made for supporting the arbor. Now file a flat into the drilled end of the rod until you have removed enough metal to cut into the hole; continue until the depth of the exposed hole is three-quarters its diameter. At this stage you can still retain the 'shelf' shown in Fig 60b, and which will assist you in filing a parallel diameter on the pivot; after a little practice, however, you will find that you do not need this assistance.

Cut off the support that you have just made and set it up as shown in Fig 61. This is the position of the arbor whether it is held with the brass support or a Jacot tool. The file is now applied to the part of the pivot that is exposed, and it will be of opposite hand to the one used before on the underside of the pivot. If you only intend to buy one file, hold it by the end and not by the handle. It is not always easy to see where the file has been removing metal from the pivot diameter; either dull the surface of the pivot by rubbing a finger on it or use a little engineer's blue, the cut surface will then show as a bright mark.

It is not absolutely necessary to remove all the lines that show on the pivot, for where a cross-section of it has 'peaks' and 'valleys', it is largely the peaks that do the damage to the pivot hole. If they are reduced to the point where there is at least 90 per cent of the original surface and the valleys that are left are too small to accept grit, there is no real necessity to go further. Indeed there is a case for leaving fine grooves of this nature as oil-grooves, but they must not appear outside the confines of the pivot hole, otherwise they will serve to transport dirt from the outside of the plate to the inside (Fig 62).

A great problem is to decide whether the pivot has had so much metal removed throughout its career that it is no longer strong enough or of sufficient bearing area. If you can see evidence of the original pivot, the

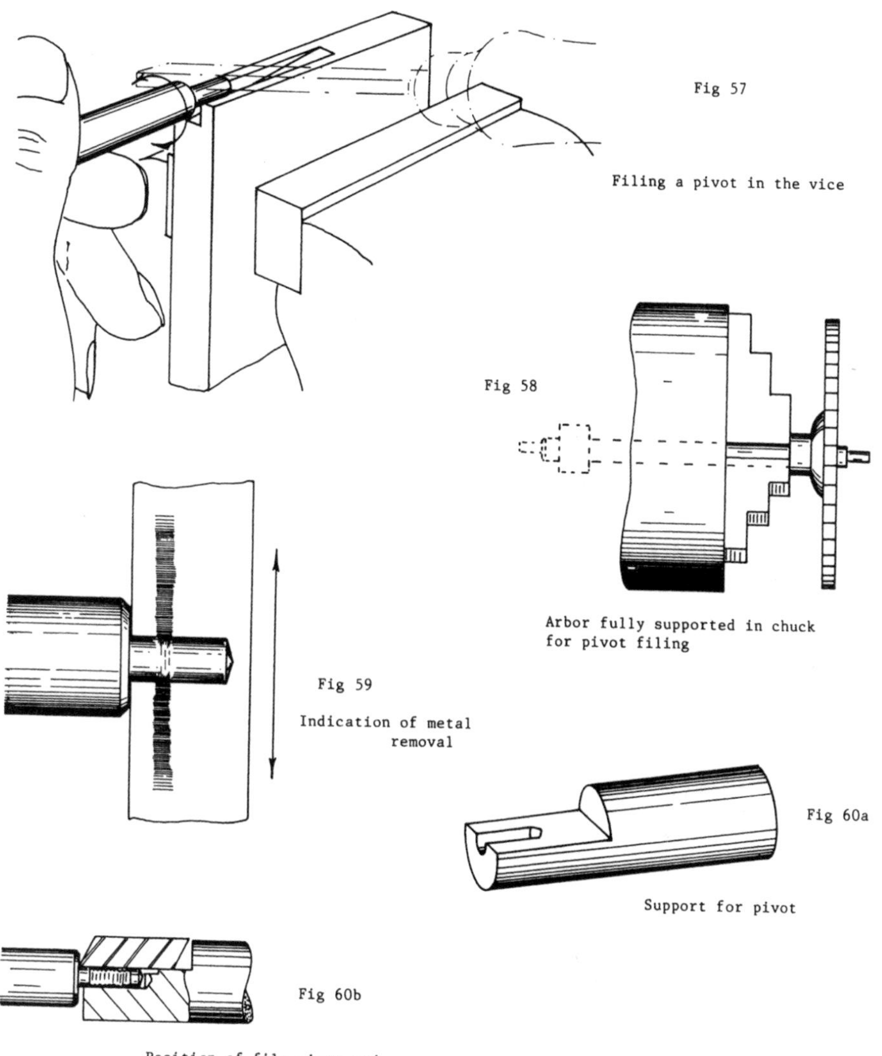

Fig 57

Filing a pivot in the vice

Fig 58

Arbor fully supported in chuck
for pivot filing

Fig 59

Indication of metal
removal

Fig 60a

Support for pivot

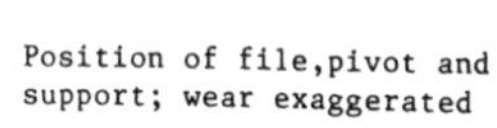

Fig 60b

Position of file,pivot and
support; wear exaggerated

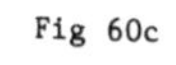

Fig 60c

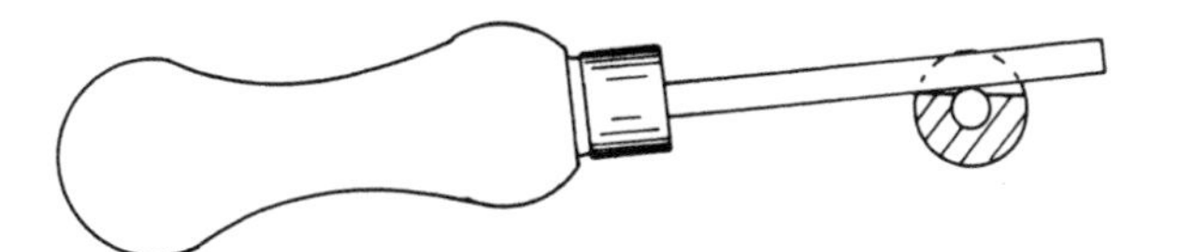

decisive point is when the pivot has reached about half its starting size, otherwise take the other pivots in the clock as an indication. An undersized pivot is more likely to wear the pivot hole out quickly than to actually break.

Burnishing

After filing it is necessary to polish the working surface of the pivot by burnishing. A pivot burnisher is a flat piece of hard steel with the same cross-section as the pivot file, but with a smooth finish all over. It has no teeth. Before use, it is placed on an emery board with emery paper of about 150 grit and rubbed from side to side so that lines are cut across its face; it is then applied to the pivot in the same manner as a file.

Use a little lubrication – a light oil or a little spit – and run the pivot at high speed, ie between 2,000 and 2,500rpm. If the burnisher is cutting well (and it does cut *very* slightly), the surface of the pivot will polish to a black shine. If the burnisher is dragging the surface, the pivot will show grey streaks or lines, in which case try a little more lubrication or a different piece of steel for the burnisher. It is well worthwhile keeping a box of burnishers, all of hard steel but from different sources, so that you can switch from one that is not working properly to one that will. Remove any sharp edges from the side of the burnisher with an Arkansas stone, to leave a hair's-breadth width of chamfer along it.

Final polishing can be carried out using diamantine mixed with oil and crushed onto a polisher, diamantine being one of the many specialist abrasives that can be bought from materials suppliers. But for the size of pivots that are used in clocks I prefer to use a piece of crocus paper backed up with the burnisher and no lubricant. This technique is not really required on ordinary longcase or wall clocks; here burnishing is sufficient. But for small French clocks, thirty-day Vienna and any other clock with very little power available to it for the duty required, the pivot diameter must be polished to as high a finish as possible.

The surface produced by a burnisher is both smoother and of different grain structure to that obtained by filing. When viewed under a microscope (after being etched slightly), steel shows a surface rather like a map of small English fields with odd shapes and varying colours – these are the grains. For our purposes the make-up of the grains is unimportant, but their different colours do reflect the different constituents of the steel. The grains, being slightly different in constituents, create small electrical currents in the presence of an electrolyte such as water as a result of an interchange of charged atoms of carbon, iron and other elements present in the metal. This causes oxides and salts to form, and this process accelerates as the electrochemical differences grow greater – in other words we have corrosion.

If the surface is very smooth, so that even at microscopic level there are no grooves to trap an electrolyte and no discernible differences in the height of the surface, the tendency to corrode is lessened. If the grains are smeared mechanically so that the difference in chemical make-up is blurred, or even totally disguised, the tendency to corrode will be even smaller. Burnishing smears the grains in just this way, and this provides a good bearing and a highly rust-resistant one.

It is often claimed that burnishing work-hardens the metal. But I find it difficult to believe that any appreciable degree of hardening can be given to steel with a hand-held burnisher used in clockmaker's fashion. The necessary deformation of the pivot metal to impart work-hardness lies outside the physical capabilities of all the clockmakers that I know. Certainly I can think of no way to measure it. Work-hardness as a result of using heavy press tools is difficult enough to gauge, that brought about by leaning on a piece of smooth steel must be almost impossible.

MAKING A NEW ARBOR AND PIVOT

If a new arbor and pivot is being made, the latter will have to be turned first and then lightly filed. It should then be hardened and tempered, and cleaned with emery paper before being finally burnished; Figs 61 to 63 show machining details.

Hardening and tempering steel calls for two heat-treatment processes, one after the other. First of all the metal is heated – in the area where hardness is required – to red heat, and is then cooled quickly (quenched) by plunging into oil or water. For clockmaking purposes the former is preferable; the hardness will be slightly lower than from quenching with water, but this is of no importance in

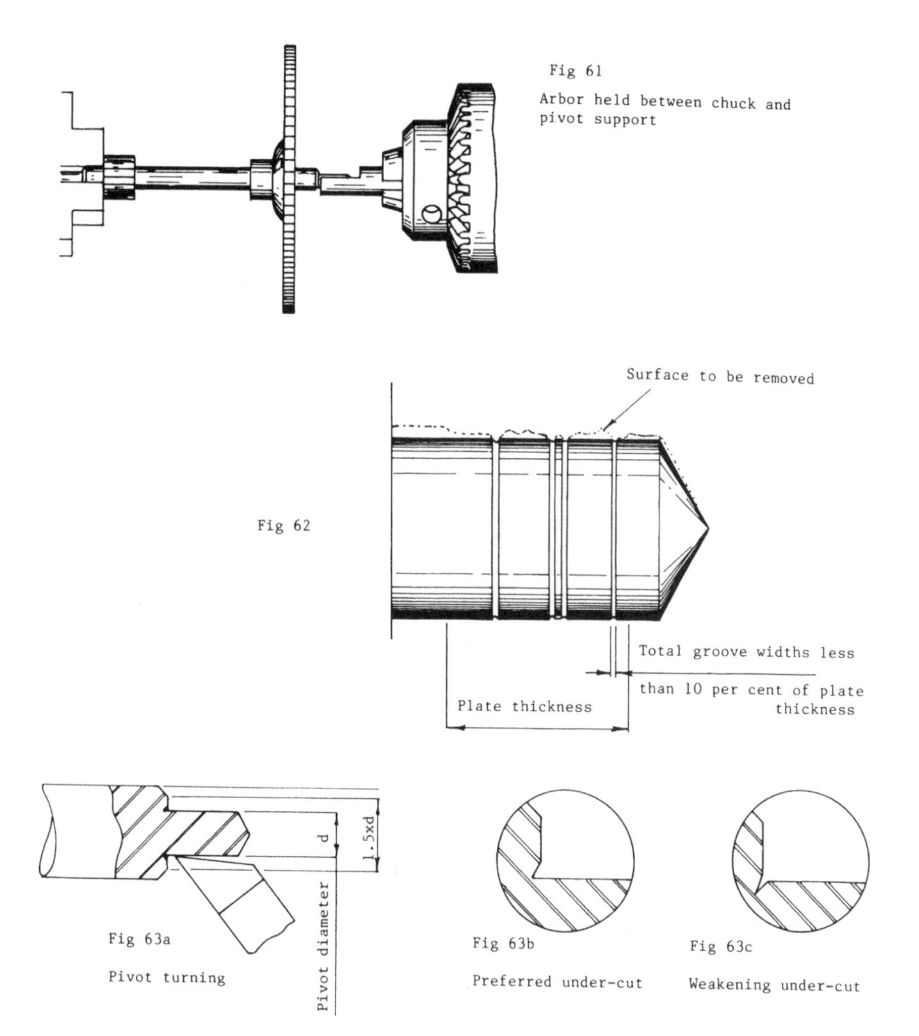

Fig 61
Arbor held between chuck and pivot support

Fig 62

Fig 63a
Pivot turning

Fig 63b
Preferred under-cut

Fig 63c
Weakening under-cut

most cases and certainly not when the steel is to be tempered anyway. Oil does not leave a rust-prone surface on the steel, and there is less risk of the metal suffering from shock at the moment of quenching.

If silver steel is used there will be little risk of distortion during quenching since it is resistant to this sort of failure; and the risk can be lessened further by ensuring that the arbor enters the oil end-on and not side-on. Other carbon steels may distort and should be wrapped in soft iron wire before heating, so that the heat enters the piece by conduction and the temperature is fairly even within the arbor. Soft iron wire can be obtained from craft shops that supply flower arrangers with materials.

The second process – tempering – is intended to reduce the degree of hardness in the metal and also make it less brittle. Pivots should be cleaned to reveal bright steel and then the arbor heated until the heat travelling along it causes the pivot to turn blue. At that point the metal is quenched again. The correct temperature can be obtained by using special salts, or an alloy of lead and tin, that melt at that temperature. The salts or metal are kept just molten, not boiling, and the parts for tempering are immersed in the liquid. However, there is a simpler method than any of these – when the steel has been quenched in oil it is immediately reheated until the oil just flames and then quenched again. This relies on using an oil for quenching that sustains a flame at blue temper, and most common lubricating oils seem to meet this criterion; but check by brightening a piece of steel and heating it until the oil flames. The temperature reached by the steel will be shown by the colour of its surface. This method is fine for the large pivots found in longcase and bracket clocks, smaller pivots of below 0.5mm (0.02in) diameter should be brightened and the tempering colour observed carefully.

RE-PIVOTING AN ARBOR

Re-pivoting is rarely a pleasure. Figs 64 to 67 show the various methods of ensuring that the drill enters the end of the arbor at its dead centre. There is a rather more expensive tool – a variation on the Jacot tool – that performs the same function, but no better or quicker than the simple method shown.

The main problem in drilling arbors for the new pivot lies in softening the pinion when it is adjacent to the pivot. If the pinion is a small one, from a carriage clock for instance, there is a danger of damaging the fine leaves when softening the steel by raising to red heat and cooling slowly. This danger can be greatly reduced by making sure that the temperature reached is only a dull red, and by dipping the pinion in a solution of copper sulphate first of all to plate it with a thin surface of copper. Earlier clocks than this have steel that is not of high quality; in particular the longcase and bracket clock have arbors of very variable quality, and you will often find there is a hard material buried in the steel just below the pivot that will defy drilling. There is nothing one can do about this unless there is space on the arbor to accept a false piece or muff (Fig 68); if not, the complete arbor and pinion will have to be replaced by a new one.

Assuming that the arbor has been softened successfully, choose a drill that is close to the diameter of the original pivot, but check your stock of pivot steel first to make sure that you have a suitable size available. (Some variation from the original diameter is acceptable to match a diameter that *is* in stock, or a piece of silver steel can be turned to the pivot diameter, notched as shown in Fig 67 and hardened and tempered.) Make the hole at least 1½ diameters (of the pivot) deep, but more if possible. A clearance between the new pivot and the hole in the arbor of about 0.025mm to 0.05mm (0.001in to 0.002in) is needed to accept the adhesive (cyano-

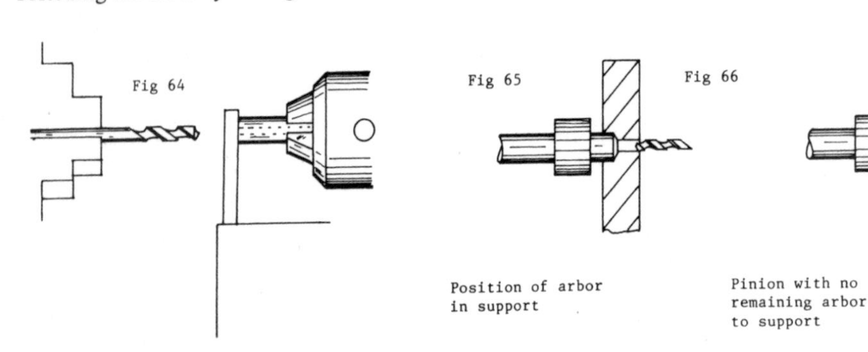

Set-up for making re-pivoting support

Position of arbor in support

Pinion with no remaining arbor to support

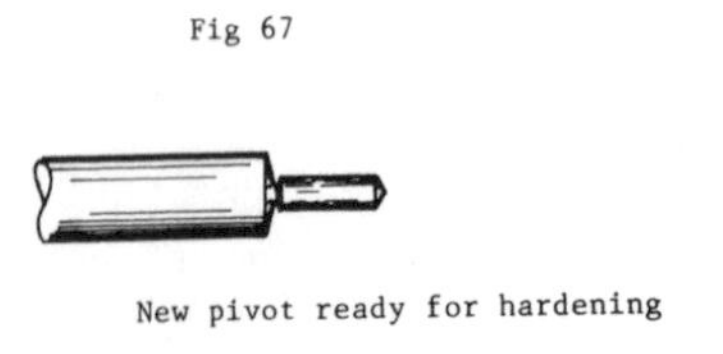

New pivot ready for hardening

acrylic). Old clockmaking techniques called for a drive fit and a tapered pivot, modern adhesives make that quite unnecessary. Before softening an arbor for pivot drilling, grip the old pivot with pliers and see if it is already a replacement, and use a small file to see whether the arbor needs softening. Many of the cheaper clocks had arbors and pivots that are soft enough to drill without heat treatment.

The end of any pivot, whether turned directly onto the arbor or inserted as above, should be finished with a cone concentric with the axis of the arbor. The conical end of the pivot is frequently used to run the arbor in a hollow runner as, for instance, when running a pinion and wheel in the depthing tool. If there is no cone, or if it is eccentric, the arbors cannot be set up properly to test the truth and mesh of the pinion/wheel pair. Old clocks often have pivots that are merely rounded with a file – persuading the arbor to run true in runners with this finish is difficult, sometimes impossible. A round end that has been turned and is, therefore, concentric, will run true, but a cone is better; it is not so adversely affected by slight damage to the end of the pivot.

Before pivoting an arbor it is often necessary to remove the wheel. Thoughtful clockmakers solder the collet that the wheel is mounted upon to the arbor, thus the collet only needs a little heat for wheel and collet to drop off quite nicely. However, some makers and repairers had a habit of driving the collet onto the arbor, and collet and wheel cannot be easily removed without risking damage to the arbor. They should be turned off in the manner shown in Fig 69, first releasing the wheel and then totally turning away the rest of the collet. If the wheel is being taken off because it is scrap, there is no problem. Take all the relevant dimensions, snip through the crossing-out to remove the rim, and then turn down the hub of the wheel and the collet. (If there is sufficient diameter of metal in the old collet to allow turning down to the diameter of the hole in the new wheel, it will not be necessary to turn it all away.) Having removed the collet and wheel, use a punch to free the wheel from its mounting (Fig 70); the collet goes into the scrap box.

To obtain true running of the wheel the bore must be concentric with the outside

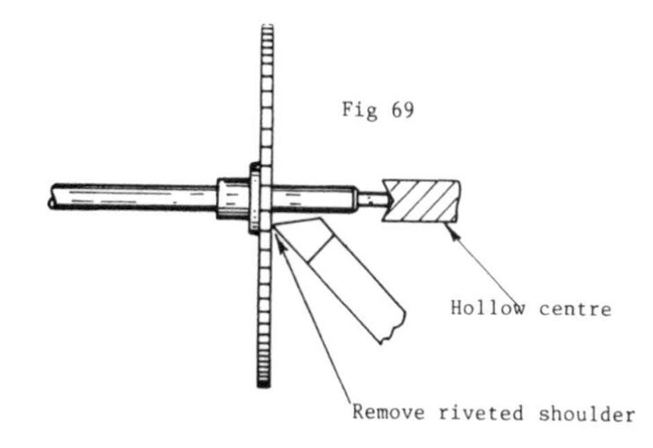

Fig 69

Removing wheel from collet

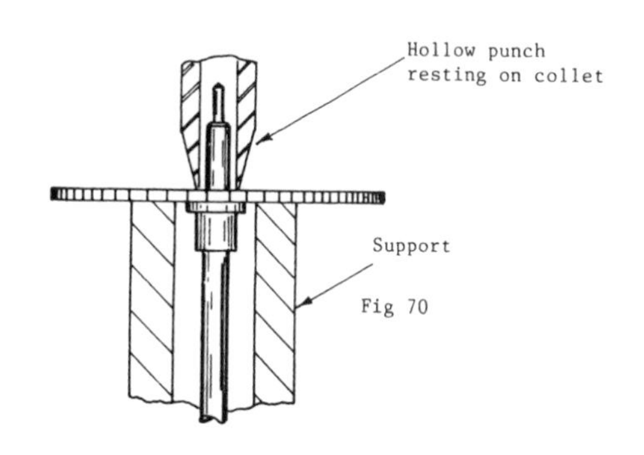

Fig 70

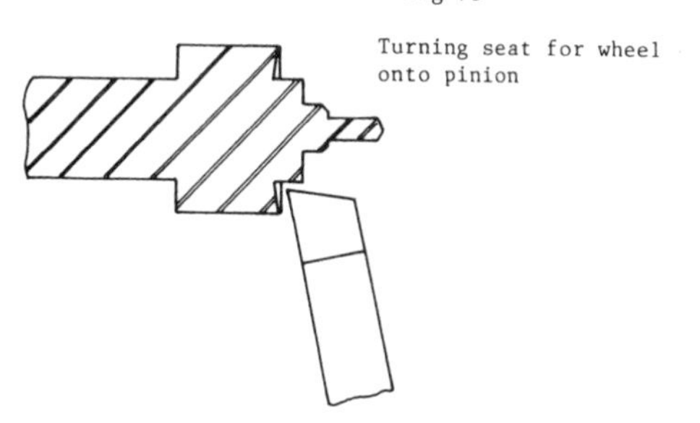
Fig 71

Turning seat for wheel onto pinion

diameter, and the seating on the collet that the wheel will slide onto must have been machined true to the pivots – hence the emphasis on concentric forms at the pivot ends. If the wheel is to go directly onto the pinion, the latter is machined to accept it (Fig 71). The hole in the wheel is commonly about 0.12mm (0.005in) smaller than the newly turned part of the pinion, so that a 'drive' fit is obtained. If too much is left on the pinion there is a danger of the wheel wandering to one side or the other as the pinion cuts its way in; a small chamfer at the beginning of the seat on the pinion will tend to prevent too 'fierce' a cutting action, but the wheel can still wander slightly. Attention must be paid to the details of the corner of the seat and the edge of the hole in the wheel so that it will sit right back against the shoulder prepared for it (Fig 71). With the wheel in position it only remains to lock it there. The one really satisfactory method is to use a ring-punch that is only 0.25mm to 0.5mm (0.01in to 0.02in) smaller than the wheel hole, support the back of the pinion and then tap its slightly protruding leaves with the punch, swaging them over.

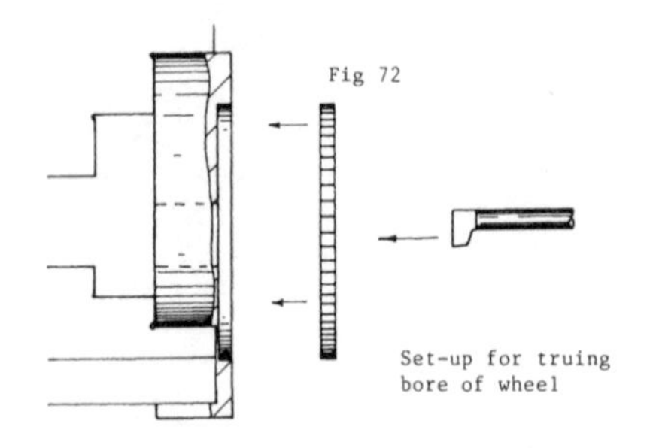
Fig 72
Set-up for truing bore of wheel

MOUNTING WHEELS

On a pinion

Mounting an old wheel on a pinion that has been re-pivoted presents a special problem in that it is often impossible to match the old bore and seating and get a concentric mounting for the wheel because grooves were cut by the original mounting and metal thrown up to one side. Take an old piece of brass sufficiently large and thick for you to machine a register into the face of the wheel's diameter about 1.5mm (0.06in) deep. Hold it in the outside jaws of the lathe chuck, and make the register (Fig 72). Traditionally, clockmakers used shellac to stick the wheel into position. The register (wax chuck) was warmed and the shellac rubbed on it until a surface of melted shellac was achieved, the wheel was then fitted in and the machinist waited until the shellac solidified. An alternative is to use a cyano-acrylic adhesive (Loctite or similar), and warm the register gently to accelerate setting. I think the latter is more positive, but there is not a lot in it. Once set, use a boring tool to machine the old bore true with the outside of the wheel, taking off as little as possible.

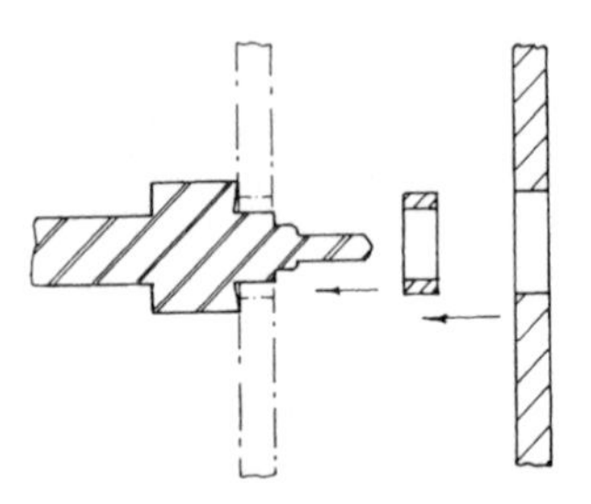
Fig 73
Bored wheel, bush and pinion

To release the wheel either warm up until the shellac melts or heat until the cyano-acrylic smokes at 150°C (300°F) approximately, and tap the wheel out onto the bench. Shellac cleans off with methylated spirit; the modern adhesive does so with a brass-wire brush.

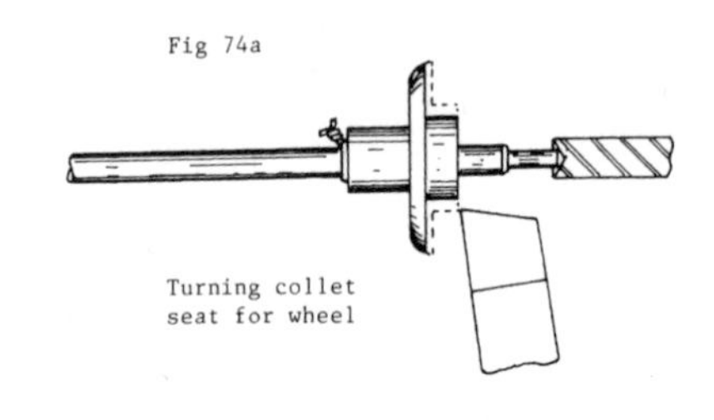
Fig 74a
Turning collet seat for wheel

The bore of the wheel is now too large to fit the pinion. Make a bush to be a tight fit on the pinion, set the latter in the chuck and make sure that it runs true, then turn the bush until the old wheel slides onto it. Fasten in place

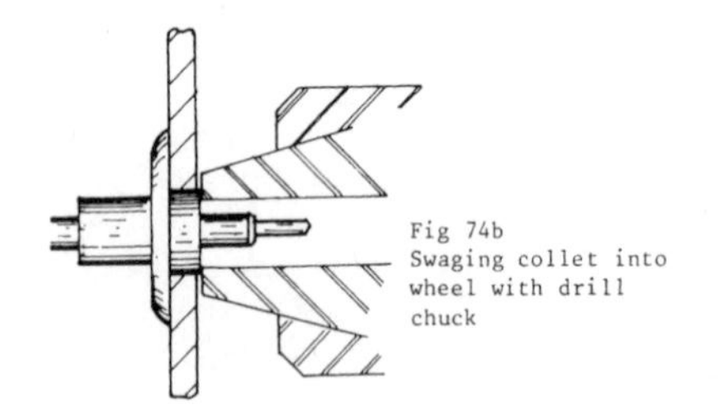
Fig 74b
Swaging collet into wheel with drill chuck

with cyano-acrylic, running the lathe at low speed and guiding the wheel with a finger-tip so that it sets to run without wobble (Fig 73).

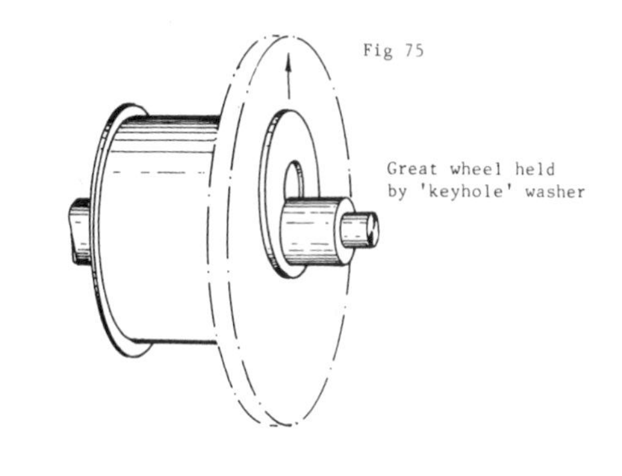
Fig 75

Great wheel held by 'keyhole' washer

Collet mounting

Collet mounting is usually a simple matter of soldering a roughly turned brass collet into a predetermined position on the arbor (Fig 74a). (A twist of fine wire will mark the position and stop the collet sliding away.) The arbor is then set up – using a hollow runner on the pivot if the arbor is tapered or ill formed, or holding the arbor directly in the chuck or lathe collet – so that the pivot runs true. When it is certain that the pivot is running true, a diameter and shoulder can be turned to seat the wheel. A nice slide-fit is required, so that the wheel just pushes on without strain and without any sideways shake. Once in position clench the wheel firmly by swaging the end of the collet as shown in Fig 74b or use a ring-punch again. The former method is best; it is very easy to ruin the setting of the wheel upon the collet by hitting the ring-punch out of square. When the mounting is complete, the wheel must run without wobble in any plane when spun between hollow runners bearing on the pivots.

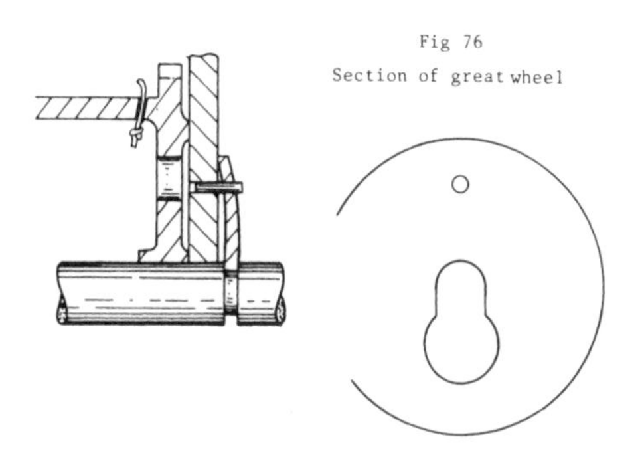
Fig 76

Section of great wheel

OTHER METHODS

There are other methods of mounting wheels on arbors. The commonest is to use a washer with a 'keyhole' in the centre and an arbor with a machined groove (Figs 75, 76). This allows the great wheel to rotate about a barrel arbor so that a ratchet and click permit winding. The wheel is free on the arbor and bears against the end of the barrel, while the large diameter of the keyhole slides over the diameter of the arbor and brings the washer up against the wheel. At this position there is a groove in the arbor, so that, by sliding the washer edgewise, the small diameter of the keyhole slides into the groove and prevents the wheel from coming off the arbor. A small pin holds washer and wheel together so that the former cannot slide out of the groove again. The washer is slightly hollowed so that it can be made to spring slightly against the wheel allowing the groove to be placed a little closer, and ensuring positive pressure against the wheel when the washer is locked.

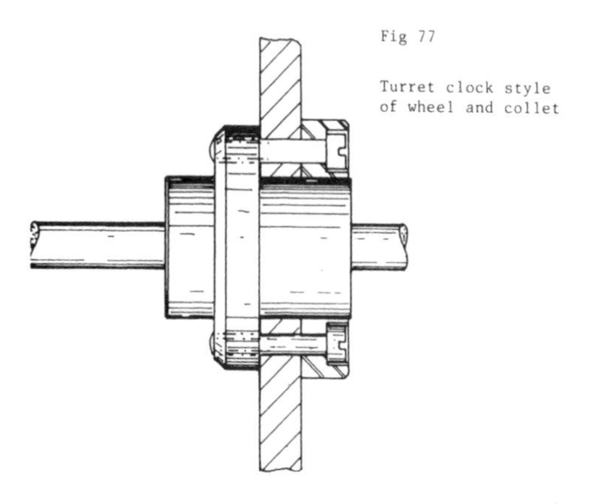
Fig 77

Turret clock style of wheel and collet

This is a very useful method of fixing the great wheel in position, but the pin that locks wheel and washer can be troublesome. Often it is left standing about 1.5mm (.06in) proud with the intention that it should be pulled out by being gripped with a pair of pliers. This is not always successful, and after two or three abortive attempts to remove the pin there will be nothing left to grab hold of. However, were there a hole in the end of the barrel that could be lined up under the pin, the latter could be driven through quite readily. Often there *is* such a hole so that the line can be knotted and the knot pulled inside the barrel, but more often than not it does not correspond with the position of the pin. Another hole, diametrically opposite the knot-hole should be put in for the next repairer to use. Some barrels have sufficient space under the great wheel to allow the pin to drive through anyway – but not many.

Large clock wheels are sometimes fastened to the collet with screws; most tower-clock wheels are fastened in this fashion (Fig 77). The only problem that this poses is that, in re-tapping a hole that has stripped its thread, the repairer will need a long-reach tap wrench. A variation on this is clear holes through wheel and collet and fitted bolts (nut and screw) holding them together.

5
Anchor Escapements

REPLACEMENT

Action of the anchor

The anchor escapement is a very reliable and long-lived form of escapement, capable of operating under the most adverse conditions. Often badly made in the first place, even more often 'repaired' by bending the pallets down to engage the escape wheel more deeply, it nevertheless contrives to keep operating until, eventually, wear makes it fail. It does, in fact, suffer more wearing of the working faces than the dead-beat escapement, because it recoils.

Recoil is the name given to the backwards motion of the escape wheel after every forward movement. It is, obviously, not as great as the forward motion, otherwise the clock would not 'go'; however, it is frequently a large fraction of that forward movement. The seconds hand of many longcase, or grandfather, clocks shows this effect quite clearly. It comes about because the impulse face (the surface the teeth of the wheel rub against) that has just reached the working position – in this case, the right-hand exit pallet – does not react immediately to the pressure of the wheel tooth and move upwards. It is 'carried over' by the momentum of the pendulum, presses down on the tooth and forces the wheel to turn anti-clockwise until the momentum is spent and the wheel reverts to its normal condition of driving the pendulum through the escapement pallets.

The amount of recoil is directly related to the wear that will occur on the impulse faces. The wheel tooth is rubbing against the face during the recoil and the normal working stroke, which increases the wear anyway; but also the impulse face is still moving towards the tooth tip at the time that the free movement (drop) of that tip takes place. The velocity of the tip must therefore be added to the velocity of the pallet, and the sum is the impact velocity. A small pit develops on the impulse face at the point of impact or first contact of tooth on pallet, caused mainly by the energy of impact. Without going into mathematical detail, it can be stated that this energy is in direct ratio to the square of the impact velocity. If the velocity of impact of one escapement is 305mm (1ft) per second and that of a similar escapement is 610mm (2ft) per second, the energy of impact in the latter case is four times that of the former. It is important, then, to keep these velocities as low as possible.

If the driving weight, or spring, is kept to the lowest practical value, two things result. Firstly, the speed with which the escape-wheel tooth approaches the pallet is kept to a minimum and, secondly, the pendulum is not driven higher in its swing than is necessary to clear the escape-wheel tooth and leave an unavoidable residue of momentum to create recoil. Wear is kept to a low level. Conversely, if the anchor is not made accurately, so that the drop on one pallet is greater than on the other, the wear on one pallet is bound to be greater than a minimum. If the effective impulse angles are not the same on each pallet, the pressure of pallet on tooth will not be the same and, even if one pallet experiences minimum wear, the other, most definitely, will not. Therefore, reliable as the anchor is, and tolerant of bad workmanship and thoughtless treatment, it will last much longer if made accurately. In addition, the practice of applying a heavier weight or stronger spring when a worn escapement increases the frictional losses in a movement will result in bad timekeeping as well as accelerated wear.

Drawing the pallets

The method of marking out pallets shown in Figs 78 to 82 has inaccuracies built into it, but these are smaller than the aggregate of errors that will result from normal marking out anyway. For instance, if a good steel rule is used, it is possible to place two lines on card to an accuracy of ± 0.1mm (0.004in), and this can be improved slightly if a sharp steel point or scriber is used. Positioning centre dots can be done to the same degree of accuracy, and both these errors will occur at each stage of marking out so that it is reasonable to claim that no system of drawing the escape pallets can be more accurate than about ± 0.4mm (0.015in); this is greater than the inaccuracies that will result from the present method. Marking out can only be used as a guide, the final form of the pallets must be obtained by trial against the escape wheel. After all, there will also be inaccuracies in the measuring and drawing of the wheel.

Figs 78 to 82 assume that only the pallets are missing or badly damaged. Unfortunately second-hand – and antique clocks are at least second-hand – movements can easily lose the escape wheel as well and, in the hands of a really enthusiastic improver, have the pivot holes so badly punched up that the centre distances between pallets and wheel are a matter of pure guesswork. It is helpful, then, to know the usual relationship between these parts.

If the radius of the escape wheel is taken to be one unit (R), the centre distance between wheel pivots and pallet pivots is about 1.5 x R; and the span of the pallets (the distance between their points) is about one quarter of the wheel's circumference. This distance varies according to the whim and convenience of the original maker, but must always be something and a half tooth spaces or pitches. A 30-tooth escape wheel has a quarter circumference of 7½, but a 32-tooth wheel has one of 8; in order to obtain the necessary half tooth the clockmaker would have chosen a span of 7½ or 8½ pitches. The insistence on a half tooth is necessary because, as you will see from Fig 82, the pallets release half a tooth at every beat of the pendulum.

If the escape wheel is missing and not simply damaged, it will be necessary to calculate the number of teeth. Measure the length of the pendulum from the underside of the suspension cock to the middle of the bob, and from this determine how many beats the clock makes each hour.

$$b = \frac{3{,}600}{\pi\sqrt{\frac{L}{g}}}$$

b = beats per hour
L = pendulum length (mm)
g = gravity 9815 $\frac{mm}{sec^2}$
π 3.142

The number of teeth on the escape wheel can be found by:

$$\frac{\text{beat x number of third-pinion leaves x escape pinion}}{\text{2 (number of centre-wheel teeth x third-wheel teeth)}}$$

The diameter of the escape wheel can be found by measuring the distance between the pallet centre and the escape-wheel centre or, as you will remember, this distance was 1.5 times the radius of the wheel. With that and the number of teeth known, the drawing in Fig 80 can be made and the same process for marking out an anchor pallet followed as before.

The figure of 1.5 is a useful average which will suit most escapements (if the escape wheel is missing also), but many British clockmakers used the square root of 2 (1.414) and American and Continental ones use figures above and below this. Any need to use one of these will be obvious when you make your drawing, because the variation was usually forced upon the maker by the layout and mechanical details of the movement.

The innermost circle shown in Fig 82 is 0.75 x R, which will give an impulse angle (the angle between the impulse face and a tangent at the tooth tip at the moment that the tooth touches the face) of close to 45 degrees. But if for some reason you wish to alter this, the circle is calculated thus:

$$I = R \times \cos i$$

I = radius of inner circle
i = impulse angle required

The other important faces, ie the drop-off faces where the escape wheel reaches the end of the impulse face, are shown as radial lines. It is not absolutely necessary for them to be radial, but it is easier from the point of view of marking out and, when the pallets are being finished by removing small amounts of metal from one face or the other, you will find that results are easier to forecast.

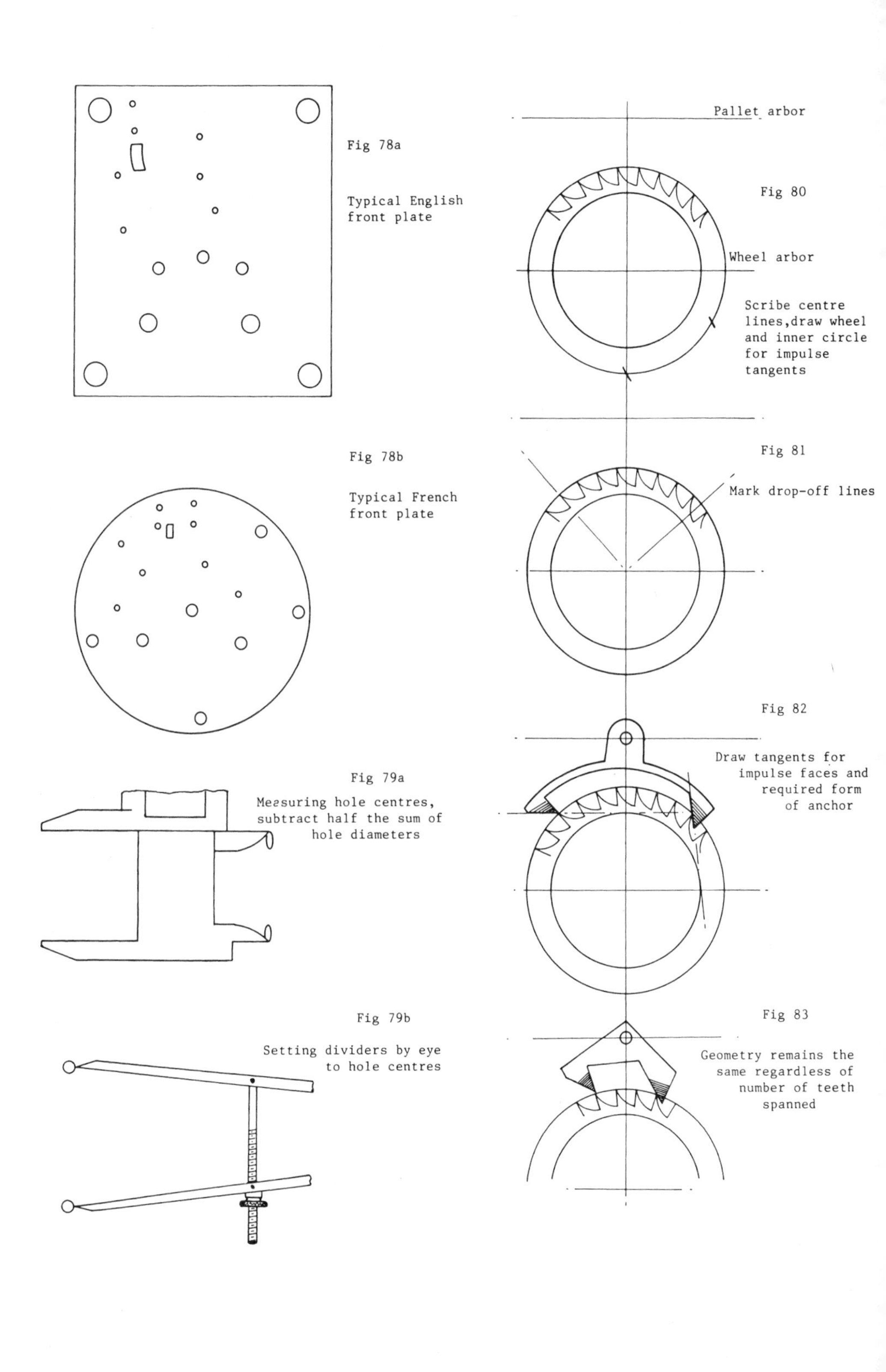
Fig 78a
Typical English front plate
Fig 78b
Typical French front plate
Fig 79a
Measuring hole centres, subtract half the sum of hole diameters
Fig 79b
Setting dividers by eye to hole centres
Pallet arbor
Fig 80
Wheel arbor
Scribe centre lines,draw wheel and inner circle for impulse tangents
Fig 81
Mark drop-off lines
Fig 82
Draw tangents for impulse faces and required form of anchor
Fig 83
Geometry remains the same regardless of number of teeth spanned

This system of drawing the impulse faces by making tangents to the inner circle from the tooth tip (Fig 82), will ensure that both pallets (entry and exit) will have the same impulse angle regardless of variations in the number of teeth on the escape wheel, number of teeth spanned by the pallets, distance between centres, or impulse angle. Therefore it can be used for British anchors, French anchors (including the tic-tac which has only one impulse face), and German and American bent-strip pallets – sometimes called verges. In fact, as is shown in Chapter 6, it will also define the impulse angles for dead-beat escapements.

In an earlier paragraph the impulse angle was defined as being measured at the moment that the escape-wheel tooth touches the impulse face; Fig 82, however, shows the entrance pallet at the top of a tooth whilst the exit is deep into the tooth space. I always draw the pallets in this way because as the entrance pallet rides up and down on the escape wheel the alteration in the aspect of the impulse face and the position of the tooth almost exactly compensate each other and the impulse angle for all practical purposes is constant. On the exit side, however, the rise of the pallet and the clockwise movement of the tooth result in an impulse angle that becomes smaller as impulse proceeds. Drawing the pallets as shown in Figs 82 and 83 ensures that the angles are equal at the beginning of impulse; if you feel they should be equal at the middle of impulse, the marking out must be made showing the escape-wheel tooth half-way along the exit impulse face. The reason for considerations of this kind on the part of clockmakers is that the escapement will perform better if the amount of energy being given to the pendulum by the pallets is more or less equal on each beat, and this is varied by inequalities between the impulse on each face and the distance to each face from the pallet arbor centre. The matter can become complex, but the simple method shown will be quite good enough for normal domestic clocks.

Many older clocks had curved impulse faces. There is no advantage to be gained from this, unless the curve results in a sudden change of angle at the point that the escape-wheel tooth first strikes. A convex curve will exaggerate the changes in impulse angle due to movement of the pallets and wheel; and if the curve matches that at the front of the tooth so that there is a face-to-face contact, it will result in the point of the pallet scraping the face of the tooth. The exception mentioned above is rather rare; it has a normal impulse angle but, at the point where contact is first made, the angle steepens so that all the recoil takes place on this steeper angle and absorbs less energy from the pendulum. It is a half-way house between the recoil and dead-beat escapement.

Alteration of impulse angle

The most common angle of impulse is 45 degrees, but clockmakers did alter this from time to time and, since these alterations affected the operation of the clock, you should try to follow the original design if it is still apparent. The reasons for having impulse angles greater or smaller than 45 degrees are as follows. The shallower the angle of impulse, the easier it is for the escape-wheel tooth to lift the impulse face; but this also results in the pallets not moving so far in allowing the escape of the tooth and, when recoil occurs, there is a greater tendency to crush the tip of the tooth. When the angle is steeper, the escape wheel finds it harder to lift the pallets, the arc that the pallets swing through to allow the tooth to escape is greater, but the crushing effect on the tooth is lessened in recoil.

The choice of angle requires the clockmaker to make a compromise between these varying effects and, generally speaking, 45 degrees is the best compromise.

Making the pallets

Make two drawings of the pallets and escape wheel on card. One of these will be stuck to the metal that the pallets are made from; the other will remain untouched until you lay the partly finished work over it, so that the important positions can be checked against a drawing that has not been obscured by filing and holding in the vice. Escapement pallets are made from steel that can be hardened, and for anchor escapements this should either be flat-ground stock or gauge plate – both forms of high-carbon steel stocked by engineers' and model engineers' suppliers – or a piece of an old file. The first two are supplied in a relatively soft state so that they can be drilled and sawn without further ado, but an old file must be annealed before work can commence. Heat the file to red heat and maintain it at that temperature for two or three minutes to ensure that it is of even temperature right through its section, then take two or three minutes more to withdraw it from the flame and cool to black. At this stage you may either plunge it into dry chalk to cool slowly or carry on withdrawing it from the flame to achieve the same effect. The type of file that I usually make pallets out of is a flat one 250mm to 300mm (10in to 12in) long; such a file produces material that is at the least equal to gauge-plate quality (as long as it was a good file to begin with).

Clean the surface of the steel – if it is an old file, remove the teeth to give a fairly smooth surface – and stick one of your card drawings onto it with a contact adhesive. When the card is firmly glued down, mark the arbor hole with a centre punch and drill it out. The outline of the anchor can be obtained by simply sawing a series of straight cuts to remove the waste, or by drilling a chain of small holes around the outside of the drawing and then sawing the links between each hole. If you find it difficult to drill the file material, you will have to re-anneal, removing the drawing as carefully as possible first.

After roughly shaping the anchor in this way, concentrate on getting the impulse and drop-off faces right, filing close to the card-outline glued to it and then laying the steel over the other card, with the arbor centre showing through the drilled hole and dead centre, and finally making the surfaces match the drawing.

Because there was no allowance made for the pallets to clear the escape-wheel teeth as they operated, there is still more metal to be removed. Mount the anchor on its arbor temporarily, and assemble it and the escape wheel in the clock plates or a depthing tool that has been set to the centre distance of the two arbors. Refer now to Figs 84 to 89. The first shows the relationship that is being aimed for, ie one drop-off point on the top of a tooth and the other mid-way between two teeth with a space between the oncoming tooth and the impulse face (the drop). It illustrates impulse about to take place on the exit pallet; precisely the same drop, with the drop-off reversed (entry drop-off mid-way between teeth and exit on the top) should obtain when about to impulse on the entry pallet. To test how close the pallets are to this state, raise one until the drop-off point is level with the top of an escape-wheel tooth as shown in Fig 86 or 87. The former shows that the drop-off points are both in correct position, with a small gap between the tooth and the impulse face on the entrance side. Therefore three faces are correct – both drop-offs

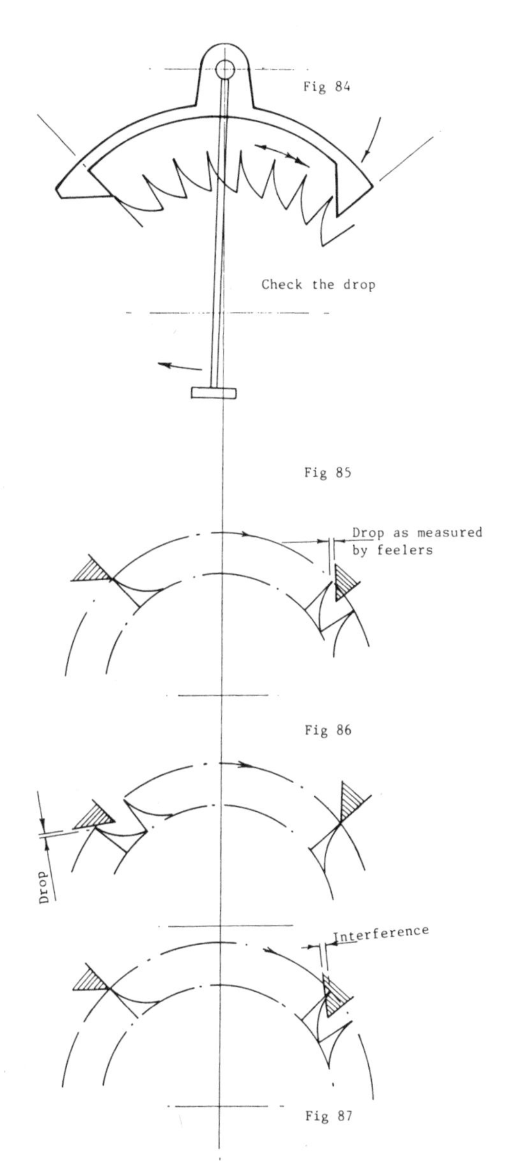

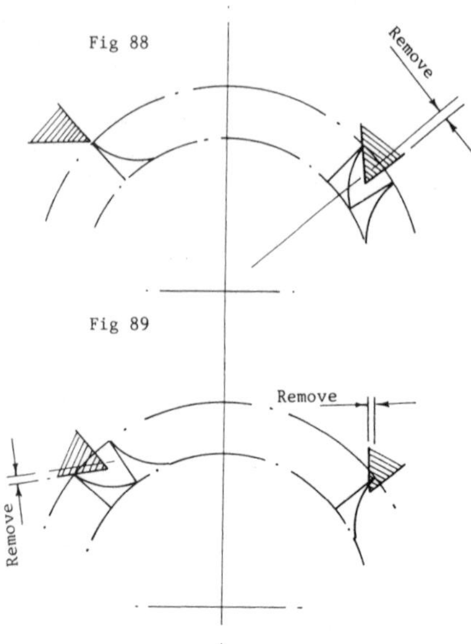

Adjusting positions of drop-off points by trial against the escapewheel

and the entrance impulse face. The latter shows that there is still too much metal on the exit impulse face.

Fig 88 has no clearance and consequently no drop on the exit face, and it has one or other drop-off face out of position. If the exit drop-off was filed back as shown, there would still be no drop on the exit. Filing metal off the exit impulse to give drop may well make everything right but not necessarily so, and before removing either the metal at the drop-off or the impulse you must try the pallets with the exit pallet at the top of a tooth. Never make a decision on the basis of one trial of the pallets, always try the escapement on the other beat and, if necessary, do so several times until you are certain where all the faces will be when you have altered any of them. Even when you are sure, only remove half the metal that is apparently due to come off, it is far quicker to make another trial than to have to go back to the beginning and cut out another pair of pallets.

Fig 89 shows pallets with both drop-off faces in the right position relative to one another. The entrance impulse should be filed until the drop-off faces move back to the radial marks, and when this has been done the exit impulse face should be filed until the exit drop-off is at the top of the tooth. All faces will now be theoretically correct but, because of the lack of drop, they will not work. File back either *both* impulse faces or *both* drop-off faces until the escapement operates without hitch; the amount removed must be the same on entrance and exit. Do not forget that the impulse surfaces must be finished to a high degree of smoothness – the pallets will look better if the drop-off faces are highly finished too – using emery paper on a board and crocus paper. Harden the impulse faces by raising the ends of the anchor to red heat and quenching in oil; do not temper, but avoid raising any more than the working surfaces to red heat. After hardening, clean all surfaces and finish impulse and drop-off faces with backed crocus paper or an Arkansas stone. The anchor can be fixed into position on its arbor by soldering, the position being set with the aid of a piece of twisted wire (Fig 90).

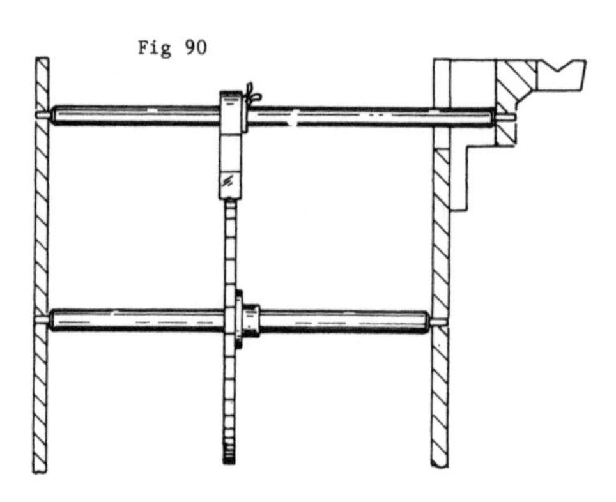

Positioning pallets ready for soldering

REPAIRING AN ANCHOR ESCAPEMENT

A worn pair of pallets will show pits on the impulse faces (Fig 91). Anneal the pallets and then replace them in the movement plates so that you can judge how much metal needs to be removed from each impulse face to give an even, though excessive, drop. Make sure, at the same time, that the pallets have their drop-off in the correct positions by going through the testing method already described; it sometimes happens that pallets have been so badly mangled by successive repairers that they cannot be repaired without

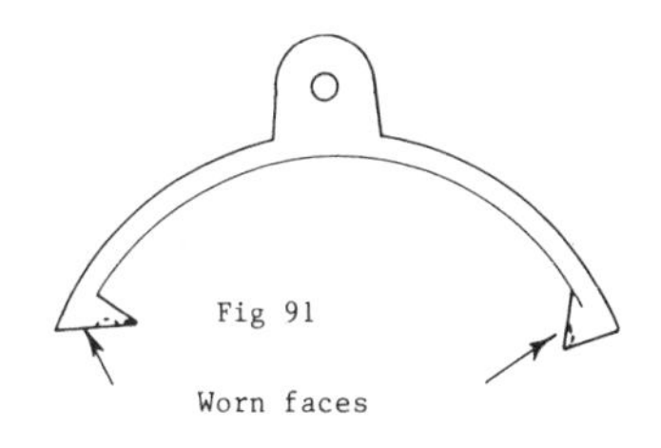

Worn faces

Fig 92

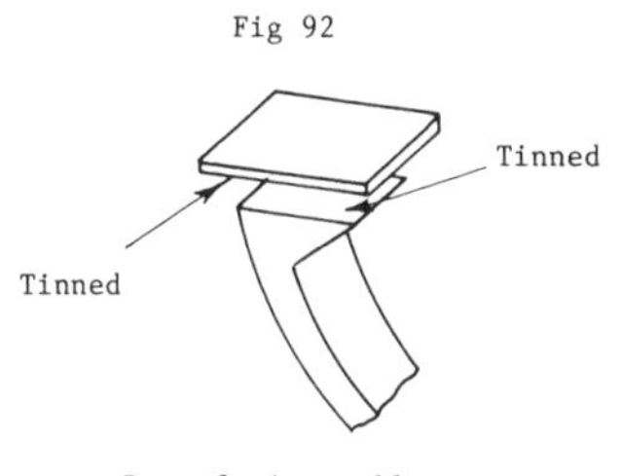

Resurfacing pallets

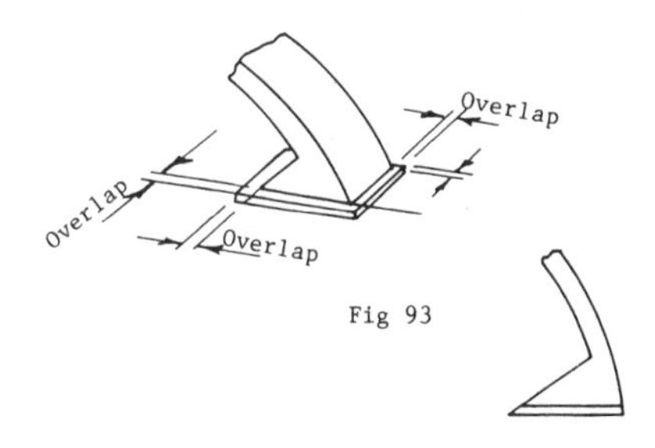

Fig 93

going to more trouble than simply making from new.

Having filed back the impulse faces, use a set of feelers to see how much metal must be soldered back to give correct drop (Fig 8 in Chapter 1). You will now be able to select a correct thickness of spring steel from your scrap-box, or file back the pallet until it will accept a piece from the box. Cut the spring steel to rough dimensions, clean and tin it with molten solder (Fig 92) using a strong flux such as Baker's fluid or killed spirit to obtain a completely wetted steel surface. While the solder is still molten, wipe it with a cotton cloth so that a thin, even layer of solder is left. Now the anchor can be tinned, also, and rested on the spring steel (Fig 93) ready for soldering. A little spot more of flux and gentle heating by gas flame should result in the two tinned surfaces melting together and showing a bright fillet around the edge of the pallet face. If a fillet does not show, but the solder is melting, hammer a piece of solder to produce a thin edge and lightly touch that to the joint until a fillet is achieved all round. A joint with the barest amount of solder necessary will make a stronger joint than one with a thick sandwich of solder. Do not put in more heat than is necessary, or the spring steel is likely to be tempered beyond blue (the melting point of soft solder), and not have such good wearing properties. Dress off all the excess steel afterwards by filing along the edges, not across them or you may pull the spring steel off again. Remove any sharp edges with an Arkansas stone and polish the impulse face.

Bending the arms of an old anchor, or lowering the pallet-arbor centre, is only acceptable as a repair if the anchor is then treated as a new one and the drop-off and impulse faces adjusted with a file or emery stone. If it is not adjusted, the geometry of the escapement will be completely wrong and, although the clock may work immediately after the repair, it will not function for as long as it should if done properly.

Bent-strip escapements

A well-worn bent-strip anchor is not normally capable of repair. If the pits and broken edges along the impulse face are not too deep they can be stoned smooth again and, if the strip is wide enough, the escape wheel can be moved to bear on an unworn part. However it is usually much simpler to make a new bent strip. The American term for this escapement anchor is a verge, and frequently one can buy new verges that only need to have the drop-off points defined.

Verges or bent-strip anchors can be made very simply from a piece of carbon steel. Old band-saw blades provide a useful source, for the body of the blade is flexible and can be bent to shape after heating to red – gauge plate does not bend easily through the angle required without producing fine cracks – but shear off the hard teeth first of all with tin snips.

Take the strip of steel after removing the teeth and softening, and file it to the required width. Note that German bent-strip anchors are often fitted into a dovetail in the arbor, and are slightly tapered and notched to allow

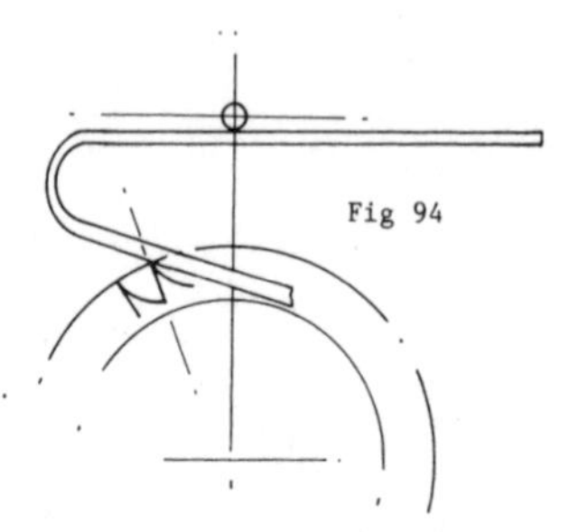

First bend on bent-strip pallets

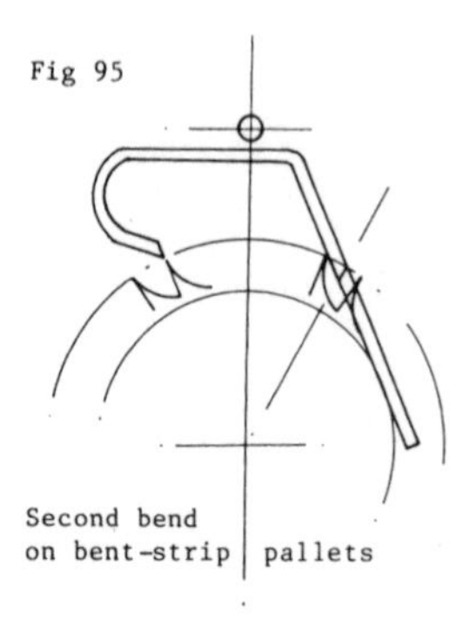

Second bend on bent-strip pallets

entry and adjustment within the dovetail.

Heat the area that is to be bent, and form the bend (Fig 94) so that the leg of the pallet is longer than necessary. The drawing of the escape-wheel teeth, arbor centres and working faces is the same as for the traditional anchor; set the centreline of the arbors to pass through the tip of a tooth if the span is an even number or through the middle of a tooth space for an odd number. Lay the strip in the position that its mounting will hold it in when complete, match the leg to the impulse angle by bending one way or the other and mark the drop-off point for cutting. Cut off the excess.

Make the second bend (Fig 95). Heat if necessary, but this bend is not usually so severe as to require heat. Match the leg to the impulse angle, and cut off at the drop-off point.

All this is done by eye, laying the work over the drawing as it proceeds. When completed, mount the pallets in the arbor and try the arbor in the clock movement; any adjustment is made by going through the procedure shown before (Figs 84 to 89) and filing back the drop-off points or bending the legs very slightly – much movement will alter the impulse angle greatly. Harden and temper the pallets to light straw to remove stresses after hardening. Polish the impulse faces and remove sharp edges.

Figs 96 to 99 show most of the different methods of mounting this design of pallet.

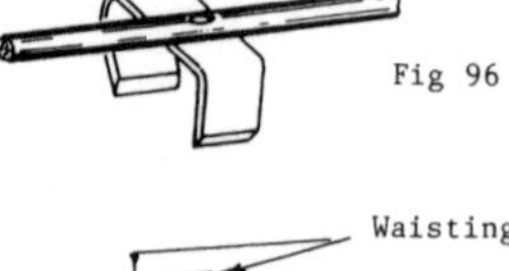

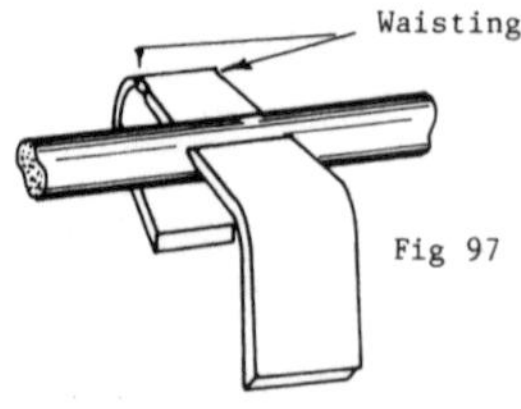

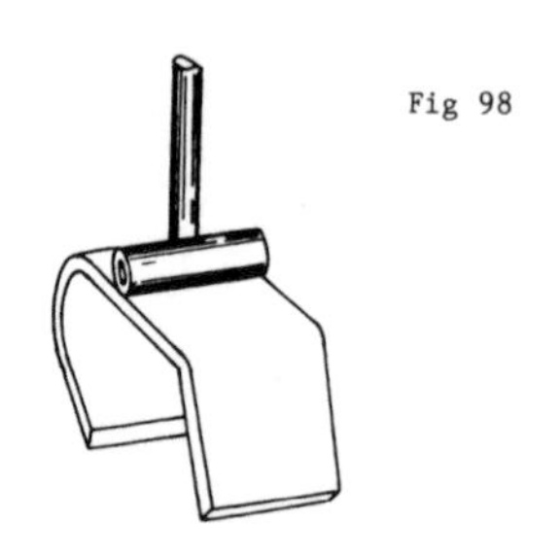

Mounting styles

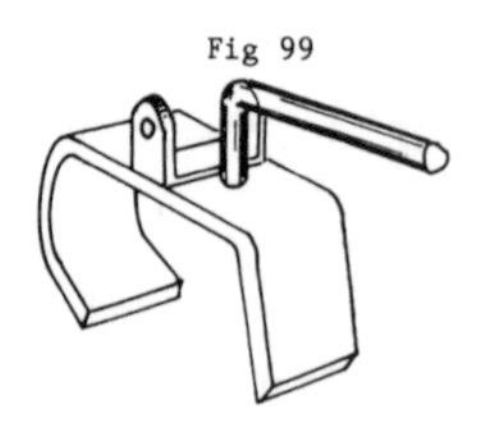

New escape wheel

The traditional escape wheel has teeth which are radially faced – often slightly raked – on one side, and concave on the other. It is this concave face that leads when the wheel is rotating, though I have never discovered a logical reason for the tradition. If you examine the geometry of the escapement you will see that it operates with no interference between wheel and pallets except at the working faces; but it *is* traditional and replacement wheels should follow the original design. Height to breadth ratios of teeth differ greatly so that some wheels have short, sturdy teeth and others long, delicate ones; and there is wide variation between these two extremes. The considerations in designing recoil teeth are:

1. The leading curve must be sufficiently steep to avoid meeting the pallet face to face.
2. The tip must be wide enough at the flat to be strong, but narrow enough to operate without a large free movement of the wheel.
3. The trailing face must not be touched by the pallet as it 'bites' into the tooth space.

The first point will normally be covered if the breadth of the tooth (the pitch), is no greater than the height. Flats at the top of the teeth should be between 0.125mm and 0.25mm (0.005in and 0.010in) for escape wheels in the range 30mm to 40mm (1.2in to 1.5in) diameters. The third point will be met satisfactorily even if the trailing face is not radial but approaching 45 degrees; and some Black Forest clocks have teeth that are of pyramid shape, sloping front and back, which makes them easy to produce by pressing from sheet brass.

Addendum

When you have made an anchor or two using the marked card it will probably occur to you that cards are not needed at all: the pallets can be marked directly from a straight edge. I believe that this is how earlier clockmakers made pallets, particularly since the inside diameter of the wheel rim (crossing-out circle) almost always gives a 45 degree impulse angle if used as the tangent circle.

Take a piece of brass plate and drill two holes at the same centre distance as the pallet and wheel arbors, to make a jig. These holes should be large enough for the arbors to pass through and for the wheel to lie flat on the plate.

Take the arbor you are using for the pallets and drill a hole for it in a piece of gauge plate large enough to make the pallets. Now assemble the steel on the arbor and pass both arbors through the holes in the plate so that the wheel lies over the steel.

Select the teeth on the wheel that will be operating on the impulse angles. Use a pencil to sketch in the entrance and exit pallets on the steel, remembering that the impulse angles make a tangent to the crossing-out circle. Position the entry pallet on the tip of a tooth, and the exit pallet sunk between two teeth, with its drop-off face on the mid-point. Twist the wheel, and modify the pencilled outline until the sketched pallets have equidistant arms. Then mark the positions of the wheel and the steel piece so you can replace them in the same relative positions.

Now mark the top of the tooth on the entry side with a centre dot and draw the radial and the impulse tangent to this point using a scriber and rule. Without moving the wheel mark the radial for the exit pallet midway between two teeth, and the impulse tangent from the tooth that acts on that pallet. Use a saw to cut away the waste within a thirty-second of an inch of the finished faces and then replace the steel.

Check that the drop-off faces are still accurate and file them to finish. Replace the steel to check the entry impulse angle, and finish with a file. Finally file the exit impulse angle. Do not file the clearance at this stage, but first check the operation of the two pallets separately on the jig. The escapement should almost work, indicating that the additional task of making the clearance would complete it. Carry out this last procedure when the pallets and wheel are mounted in the clock movement plates to allow for any errors or discrepancies in the plates. After polishing, the escapement is ready for hardening. I find that I can make a complete anchor and harden and polish it in about an hour and a quarter using this method. It is more direct than drawing on card and errors are corrected as you proceed instead of only becoming apparent when all faces are partly finished.

6
Dead-beat Escapements

REPLACEMENT

Action of the dead-beat escapement

The dead-beat escapement was developed by Graham in about 1715 to provide an escapement that avoided the problems associated with the recoil anchor. In the latter, the escape wheel is driven backwards, immediately after making contact with the impulse face, by the continuing swing of the pendulum. The dead-beat however, provides a face for the tooth to drop onto (the dead face), that makes a true radius from the arbor centre (Fig 100). Any movement of the pallet can have no effect until the impulse face is exposed; the swing of the pendulum can never cause recoil as long as the escapement is in good condition. The distance from the point on the dead face that is first contacted by the tooth to the beginning of the impulse face, is called the lock. The distance that the tooth on one side of the pallets moves to contact the dead face when the tooth on the other side comes clear of the impulse face, is called the drop.

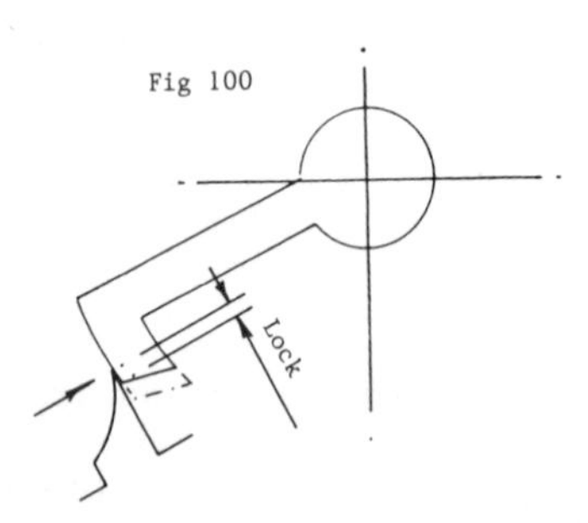

The movement of the pallets – whether caused by a pendulum or by a balance wheel – allows the escape-wheel tooth to drop onto the dead face, lock there during the supplementary swing, begin impulse at the end of that swing, and end impulse whilst the other tooth makes its drop. Any escapement that prevents the recoil action is called a dead-beat, this includes the traditional Graham, the Vienna version of the Graham, the pin-wheel, the cylinder and the Brocot.

Drawing the pallets

This drawing will not be used to form a direct guide to the shape of the pallets, but more as a matter of calculating dimensions. Fig 101 shows the normal layout of the pallets, which can be defined by determining two radii and the lock.

Set the drawing of the escape wheel so that, if the span is an even number, the centre line passes through the tip of a tooth and, if it is odd, the centre line passes through the middle of a space. The outside radius O is the dead face of the entrance pallet and the drop-off point of the exit pallet; I is the inner radius and is the dead face for the exit pallet and the drop-off for the entrance. These are the two radii that are used to make the pallets. Radius T passes through the tips of the incoming and the outgoing teeth.

The dead-beat escapement needs clearance for operation just as the recoil (drop) does; drop being often defined as an angle of rotation of the escape wheel, though I prefer to give it a dimension that can be measured by means of feelers. The drop is kept to as small an amount as will allow the reliable operation of the escapement – often about 4 to 5 per cent of the tooth pitch.

The outside radius O is equal to radius T minus half the drop:

$$O = T - \frac{drop}{2}$$

The width of the pallet (O – I) is equal to half the pitch (p) minus the drop, which is easier

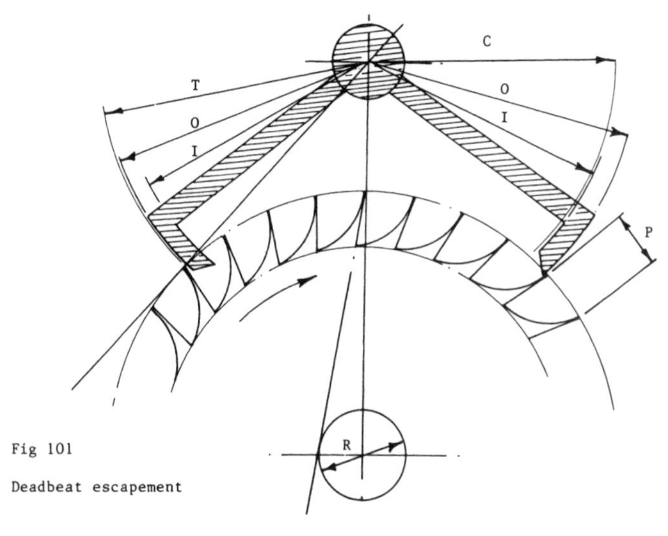

Fig 101

Deadbeat escapement

P Tooth pitch
T Position of entry wheel tooth before drop
C Position of exit wheel tooth before drop
O Outside radius = T − ½ drop
I Inside radius = C + ½ drop

T−C = ½ P

R Circle for defining rake of teeth

to understand than

$$I = T - \left(\frac{p}{2} - \frac{drop}{2}\right) = T - \left(\frac{p - drop}{2}\right)$$

Clearly the radius T, which is measured from the arbor centre to the wheel-tooth tips, can either be calculated by trigonometry or it can be measured directly by drawing out several times full size, very carefully. To obtain T by trigonometry you will need to measure the distance between arbor centres (d); the radius of the escape wheel (r); the angle between the centre line and a radial drawn to the tip of the spanned teeth = $\propto$. The angle can be found by multiplying 180 degrees by the number of teeth spanned, then dividing by total number of teeth on the wheel. In Fig 101 this is

$$180 \times \frac{8 \text{ teeth}}{28 \text{ teeth}} = 51 \text{ degrees } 26 \text{ minutes}$$

and $$T = \sqrt{d^2 + r^2 - (2.d.r \times \cos \propto)}$$

Making the pallets

The pallets can be made from gauge plate as the recoil anchor was; centre punch the arbor hole, and use dividers (engineer's compasses) to draw the outside and inside radii. Decide the approximate position of the impulse faces, leaving metal for filing off about 0.5mm (0.02in) from each. The length of the curved nibs is not at all critical, but must be enough so that there is no possibility of the radial arms striking the tops of the escape-wheel teeth. The outside radius O can be produced to close tolerances by turning on the lathe, but the inside radius I (on the traditional Graham) must be filed to a mark that is as accurately scribed as possible. Vernier callipers can be used, once O is machined, to check the width of the pallet and, thus, the inside radius I. When the arms are finished and the radii of the nibs, the impulse faces can be filed.

The impulse angle commonly used is 45 degrees and, if the original angle cannot be determined, this is usually satisfactory. Drill a piece of plate – brass or steel – with holes the size of the pallet arbor and escape-wheel arbor, and place the escape wheel and pallets flat upon it, locating the holes with short lengths of silver steel. Make a disc of brass that is bored to drop over the collet of the escape wheel and lie flat against it, and find a short piece of steel to use as a straight edge. This disc should be 0.707 times the diameter of the wheel, minus twice the straight-edge thickness. Now the straight edge is laid against the outside of this disc and will define a 45 degree impulse angle at any point of the wheel circumference. Position the straight edge to give the impulse for the entry pallet and file the pallet to match the straight edge; the device shown in Fig 102 is an adjustable variation of the above. When a full impulse face has been produced on this pallet, position it so that it gives the lock required, clamp it down and turn the wheel until a tooth comes up to rest on the locking face of the exit pallet. Use the straight edge again so that an impulse angle can be filed here also. It cannot of course be filed easily whilst clamped and it is best to mark the angle, file roughly, and then remove the last hair's breadth when the pallets are clamped on the plate – remove the wheel please.

Set wheel and pallets in the movement plates and use emery paper and a board to produce smooth impulse faces. Harden when the escape wheel turns freely with the required amount of lock on each pallet; do not temper, but polish with crocus paper or Arkansas stone.

Other impulse angles are frequently seen in deadbeat escapements. For a 35 degree impulse angle use a disc of 0.82 times the diameter of the wheel, and for 30 degrees one of 0.866 times the diameter.

REPAIR

Graham

Badly worn pallets should be replaced because the traditional Graham is a one-piece device, there is no adjustment between the two arms of the pallets. Try to reproduce the original as closely as possible because this escapement hardly ever appears on a poor movement and an obvious replacement not only detracts from the original design but will affect the value.

Slight wear, a little pocketing of the locking face or roughening of the impulse, can be corrected by the use of an Arkansas stone. There will be a larger drop or a smaller lock, but so long as the escape-wheel tooth can never drop directly onto the impulse face, or

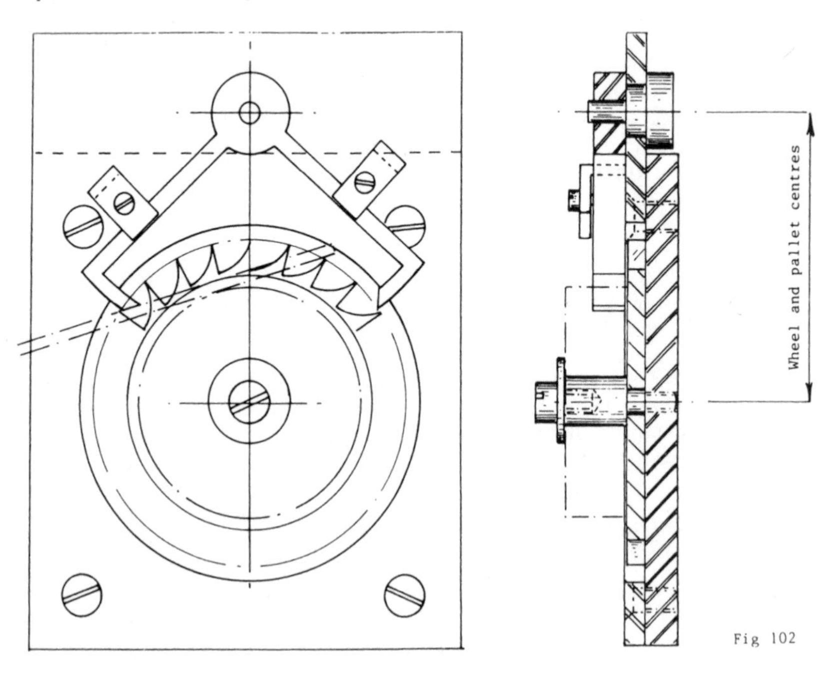

Jig for filing impulse faces. Chain dots show file and a disc of a diameter that will place the cutting surface of the file on the impulse tangent

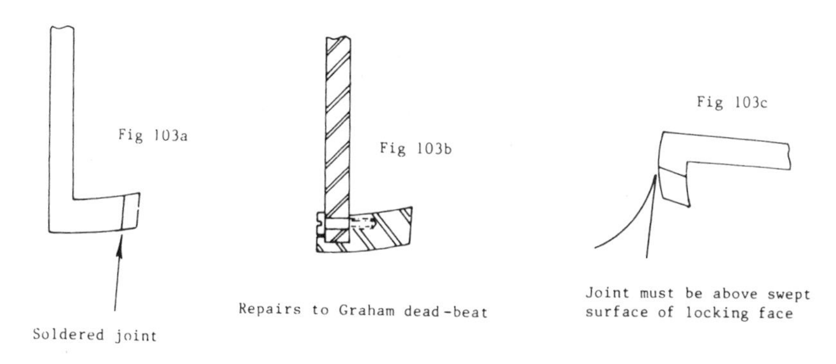

Repairs to Graham dead-beat

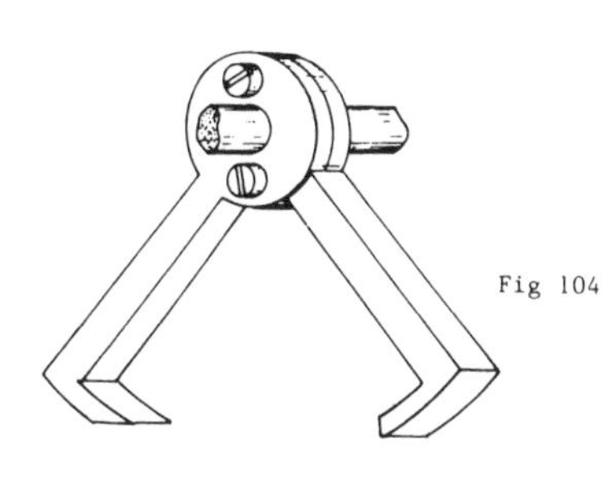
Fig 104

Split arms

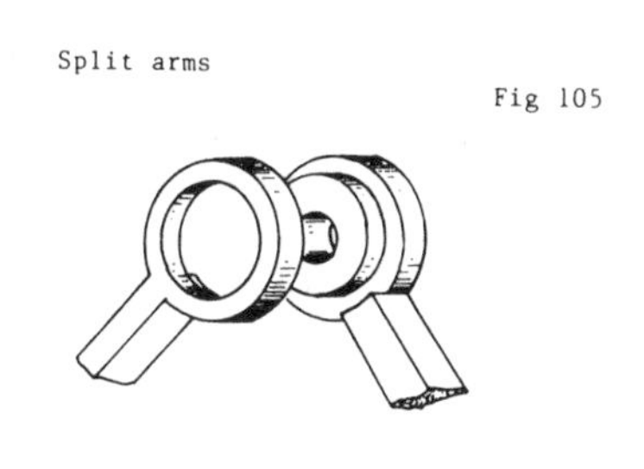
Fig 105

Fig 106

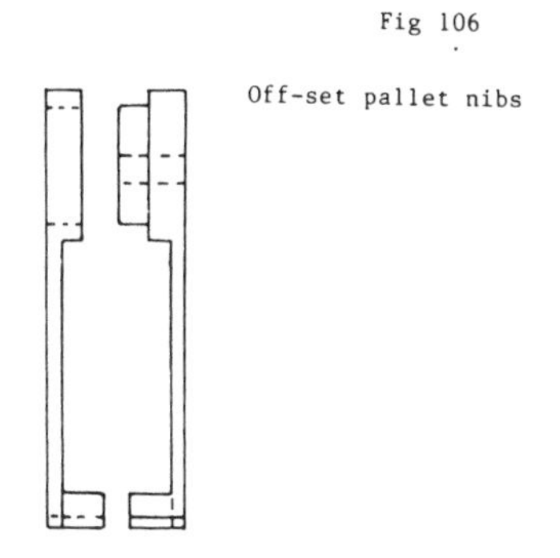
Off-set pallet nibs

even onto the corner of impulse and lock, the escapement will perform quite satisfactorily. This is an important point, common to all dead-beat escapements; under no circumstances should the wheel tooth be able to drop thus, because the movement will then have a recoil escapement and it does not have enough energy to operate as such.

If there is more than slight wear and a new escapement is to be avoided, the only alternative is to add a piece of hardened carbon steel. Figs 103a and 103b show a soldered addition and a screwed-on nib respectively. Leave sufficient metal on the addition for the outer and inner radii to be stoned to suit the escape wheel. If the soldered addition is used, the joint must be above the highest point that can be wiped by the escape-wheel tooth (Fig 103c). Figs 104 to 106 show Graham dead-beat pallets that can be made in two parts so that the positioning of the impulse faces can be adjusted; the last figure is a side elevation of an off-set pair of pallets that can be turned to produce both inner and outer radii.

Repairing two-part pallets is relatively simple. If there is only slight wear on the impulse faces, they can be stoned and left at the original setting as if they were traditional pattern. When deep pits are present, the impulse can either be stoned back or each nib can be annealed and then filed back using the techniques described for making the Graham impulses from new. The arms are then unlocked, moved until the correct lock is obtained and then drilled for pegging together before fitting screws again, after re-hardening if necessary. The pegs (or dowels) locate the parts accurately and the screws hold them firm.

Vienna type dead-beat

The only difference between the type of dead-beat escapement that is found in the Vienna regulator and the Graham is the construction of the pallets; the operation and geometry remain the same.

The escapement consists of a shaped bar that is grooved to accept pallets that are cut from a ring of carbon steel (Figs 107, 108). Since the pallets are made from a ring, the inside and outside radii can be turned very easily. The impulse angles are obtained by cutting the ring in such a way as to leave the entrance impulse angle at one end of the piece and the exit impulse at the other. When the escapement shows wear the pallets can either be stoned or, if the damage is great, they can be rotated (Fig 108) to present untouched impulse faces. The locking is adjusted by sliding the pieces along the grooves and then clamping in position with the little straps provided. If pallets are too damaged at both ends, new ones can be made easily by turning a ring from a piece of large-diameter silver-steel bar, or some other high-carbon steel. The impulse angles are obtained in the same manner as before.

The Brocot

DRAWING THE ESCAPEMENT

This escapement is very frequently found in the better French mantel clocks and many others where a good mass-produced escapement is needed. Apart from its reliable performance it is also attractive and is featured in the visible escapement clocks that have the whole assembly of wheel and pallets displayed on the dial. Fig 109 shows the important features of the escapement – the support for the pallets is not shown, since it does not affect the operation at all and varies from a plain, inverted Y to a highly decorative version.

The pallets are half cylinders of hard steel or jewel (usually a commercial garnet) set in drilled holes, so that when the correct positions of arbor and pallet holes are determined it is easy to produce them in large quantity by simple jig-drilling techniques. The impulse face on the pallet is a quarter of a circle and the locking face is more properly described as a point, since it is that part of the circumference that makes a tangent with a radial from the escape wheel. Because of this it is important that the front face of the escape-wheel tooth makes a tangent with the arc (struck from the pallet arbor) that the locking point swings through, otherwise there will be recoil.

Fig 109 has the arbor centre marked, and the position A marks the tangent to the diameter of the wheel. If the tangent coincided with tip of one tooth and the mid-point of a tooth space – the drop-off points – this would define one of the centre lines for the pallets. However, it rarely does, and the drop-off points are chosen to be as close to the tangent as convenient, with the result that the teeth must be designed (by the manufacturer) so that their front faces make tangents to the swing of the pallets at this position. This consideration will only be of importance if the whole pallet assembly is missing and you need an indication as to which of the possible positions for the pallet holes the original manufacturer selected. Place the new holes where the escape-wheel teeth will make a tangent to the swing of the pallets.

Fig 110 shows the entry pallet and the exit pallet, one of them giving the drop to the locking position and the other having just finished giving impulse. The centres for the pallet holes lie on a radial struck through the drop-off point and, in the one case, half the pallet diameter above the wheel circumference; in the other, directly on the wheel circumference. Which pallet is placed on the circle and which above does not matter, since the situation reverses as the pendulum swings. There are two ways of deciding the pallet diameter (Fig 111). The most common is to measure the wheel-tooth pitch and then subtract twice the drop that you propose to use – established by the same rule as the Graham at about 5 per cent of pitch. The pallet is then halved to produce the drop-off face. The other method is to make the pallet the same diameter as the pitch and then to produce the drop-off face below the half-cylinder by an amount equal to the drop. The advantage of the second method is that the pallet continues to give impulse right up to the drop-off. Both types of pallet adjust the drop-off slightly by twisting in their sockets (Fig 112).

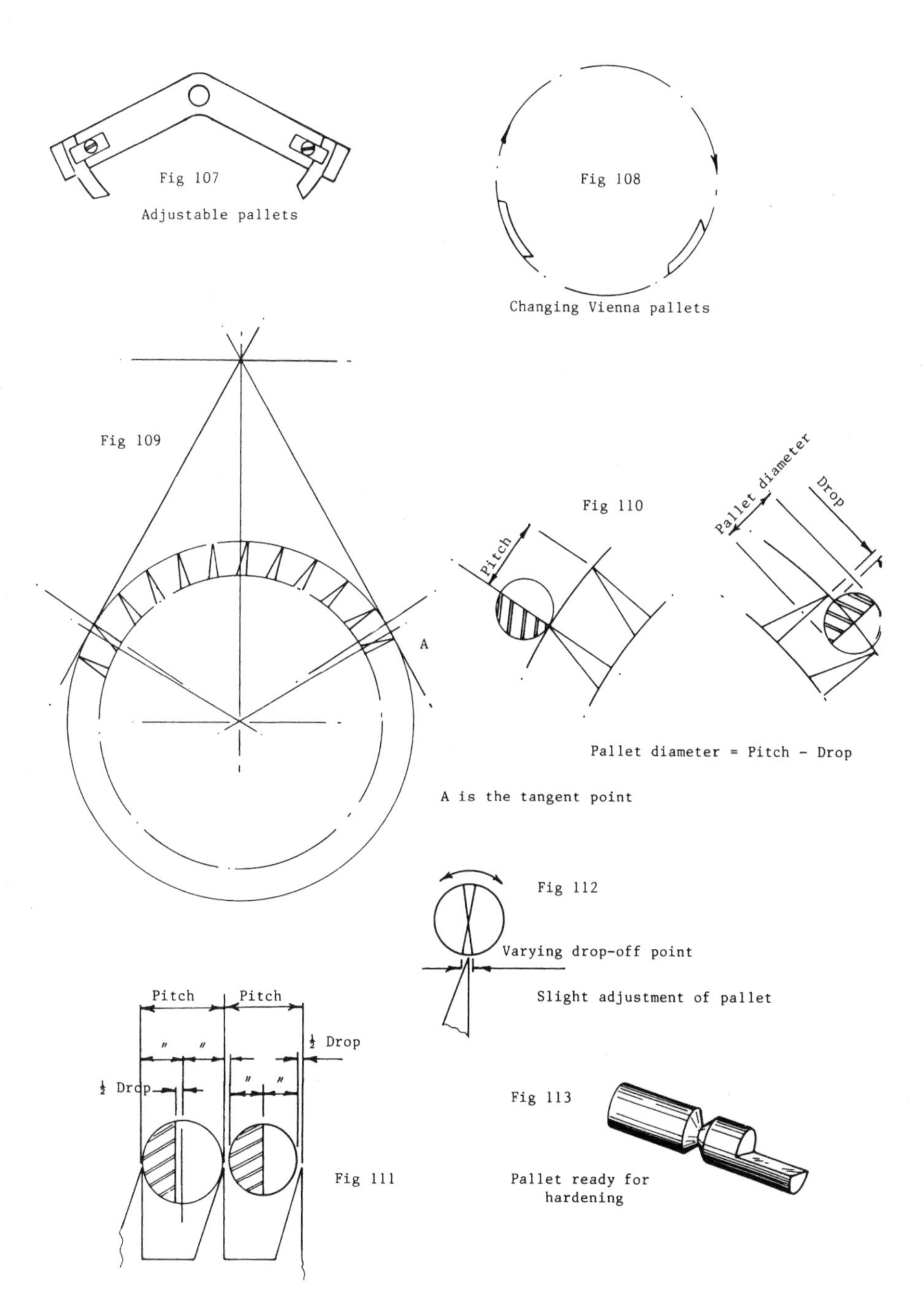

Fig 107
Adjustable pallets

Fig 108
Changing Vienna pallets

Fig 109

A is the tangent point

Fig 110

Pallet diameter = Pitch - Drop

Fig 111

Fig 112
Varying drop-off point

Slight adjustment of pallet

Fig 113

Pallet ready for hardening

MAKING THE PALLETS

The actual half-cylinders can be bought in a limited range of sizes, manufactured from steel or garnet; but they can also be made from silver steel very simply. If these are replacements for an otherwise undamaged pallet assembly, remove the old pallets (if they are there), clean the pallets and pallet holes with methylated spirit and then measure the diameter of the pallets or the hole. In the former case all the important dimensions will be clear and you can simply go ahead and make the replacement. However, some pallets were made with a diameter at the fitting into the hole that was larger than the actual working pallet. If the pallet is missing you must measure the hole and the pitch of the escape-wheel teeth. The pallets should be turned on the end of a silver-steel bar, filed until the half-way point is almost reached, notched (Fig 113, p75), then hardened. The pallet can be broken off the bar easily after hardening.

MAKING THE PALLET BODY

Unfortunately, you will often find Brocot escapements that have had the body of the pallets opened up or squeezed together by a repairer in an attempt to correct some fault in the escapement, and it is very likely that you will have to make a new body using the method already shown for marking out. The simplest way is to use a plate of brass thick enough for the body and drill the holes for pallet and wheel arbors, then take the escape wheel off its collet and fasten it to the plate. Dividers can now be used to align the wheel teeth to place the drop-off points equidistant from the pallet-arbor centre, and then proceed with the marking as described (Fig 114a). Finding the actual centres for the pallets themselves will be easier if you make a punch of the same diameter as the working diameter of the pallets, turn a short point at its centre and then use that to mark the drilled holes for the pallets (Fig 114b). The diameter of the punch simulates the circumference of the pallets; the simulation can be carried further by halving the punch, but it is hardly necessary and will weaken it.

Many Brocot bodies have decorative turning around the holes for the arbor and pallets. Figs 115a to 115c show how a dee-bit can be made by using the lathe to make a formed face and then filing back to make this a cutting edge.

The pallets should be stuck into the body with shellac; this enables the repairer to adjust or remove them, by warming the body. Modern adhesives are too positive for

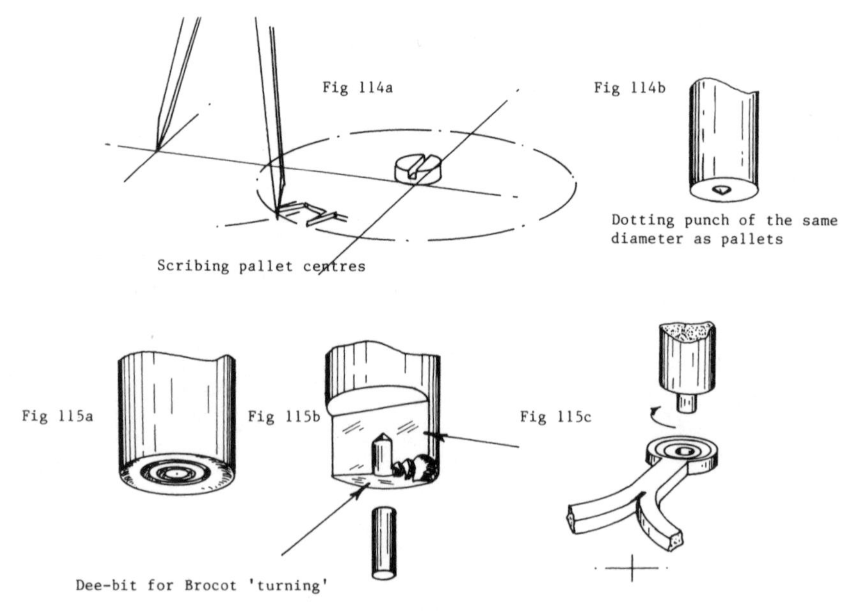
Fig 114a

Scribing pallet centres

Fig 114b

Dotting punch of the same diameter as pallets

Fig 115a

Fig 115b

Fig 115c

Dee-bit for Brocot 'turning'

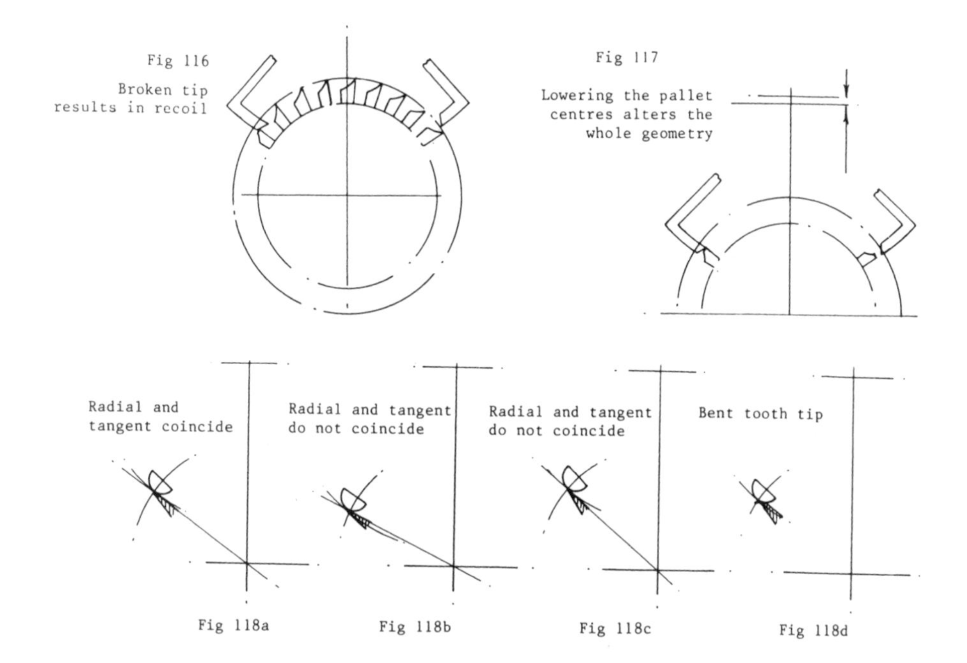

this purpose. A little clearance between the hole and the pallet will be needed for the shellac – about 0.05mm (0.002in) should be sufficient.

DEAD-BEAT ESCAPE WHEELS

Graham

The front face of the wheel tooth (the radial or raked face), should not touch the pallets; only the tip of the tooth must make contact and most dead-beat escape wheels are raked to achieve this, giving the tooth the appearance of leaning forward. If the face of the tooth does touch the locking face of the pallet it will cause a small amount of recoil. Similarly, if the wheel has a broken tooth (Fig 116), recoil will develop at that tooth. The tooth should be replaced entirely applying the same technique as that used for replacing gear teeth (Chapter 8). Lowering the pallet arbor (Fig 117) will only alter the geometry, and will not allow machining of the teeth to bring them all to the same level. Either the pallets or the wheel must be replaced and the latter is more convenient. Other faults of the wheel teeth are similar to those in the recoil escapement and may be corrected in the same way.

Brocot

Because the Brocot employs semi-cylindrical pallets, it is important that the tooth's tip strikes the dead-point of the pallet directly. In other words it must follow a path that would pass through the centre of the pallet if it could carry on. Fig 118a shows correct alignment. The tip is on the centre-line, and an arc struck from the arbor centre shows the tooth face to be tangential to it. In Fig 118b the tip is striking below the centre of the pallet and the arc cuts into the tooth face; this will give a definite recoil. Fig 118c shows the tip on centre but the tooth face not tangential to the arc; as the pallet swings into the centre the tooth will move forward slightly, so that before the pallet can swing back to give impulse there will be an equally small recoil. Fig 118d shows the result of damage to the teeth – the tip is bent back and very strong recoil will result.

Platform escapements

The platform escapement is a separately mounted escapement that is used on carriage clocks and several other types. In its most common form it consists of a balance wheel and spring and one of two types of dead-beat escapement – the cylinder and the lever.

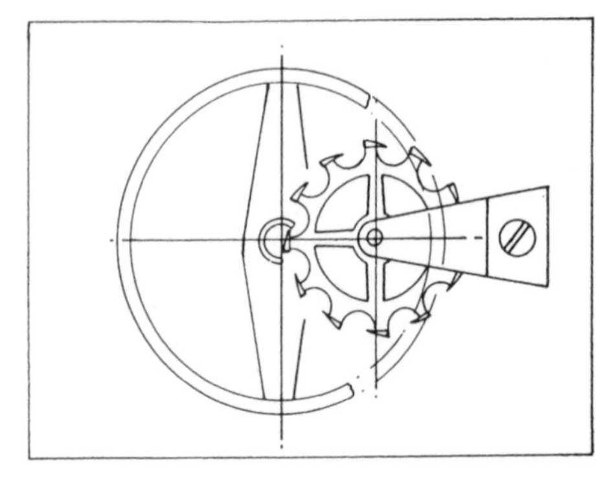

Fig 119 Cylinder escapement

Cylinder escapement (Fig 119)
The pallets of this escapement are carried on the staff or arbor of the balance wheel in the form of a cylinder. The entrance lock is made on the outside of the cylinder and the exit lock on the inside, the edges of a vertical slot serve for the pallet faces, but the impulse is given by small wedges on each tooth.

The cylinder escapement is found on the cheaper varieties of clock and is generally reckoned to be of lower quality than the lever. Replacement platforms of this type are no longer available and there are only a few craftsmen who are capable of repairing them, consequently, if the escapement is broken or so badly worn that it will no longer operate properly, it can only be replaced by a lever platform. The escapement is in beat when the same movement clockwise and counter clockwise of the balance wheel, releases the escape wheel.

Lever escapement (Fig 120)
There are several versions of the lever escapement, but they all consist of a balance wheel that carries a means of receiving impulse from a lever which also has a pair of dead-beat pallets mounted on it. The escape wheel gives impulse to the lever and the lever passes it to the balance wheel. Repair lies in the province of the watchmaker and will not be dealt with here, but the major points to inspect when judging the condition of the escapement are: the pallet stones, the other end of the lever that impulses the balance wheel, and the bearings of the staffs. The escapement is in beat when the centres of the lever and balance staffs are in line with the impulse pin (Fig 121); however, on some cheaper lever escapements the pallets are movable. In this case the centres must be lined up first by adjusting the balance wheel for beat (see next paragraph) and then adjusting the pallets. I have never seen this feature on a carriage clock, but only on vertically mounted platforms of mantel clocks.

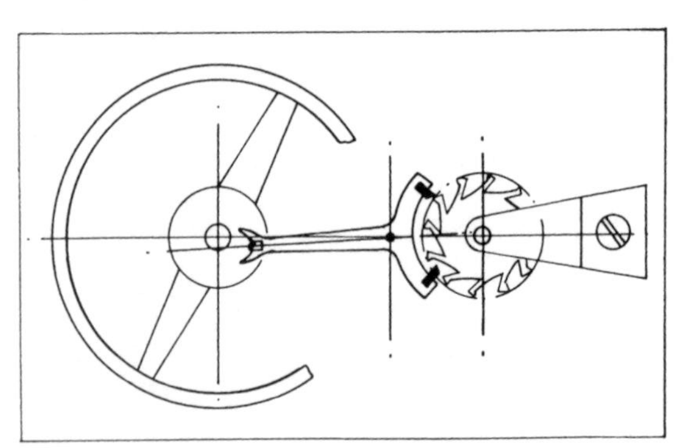

Fig 120 Lever escapement

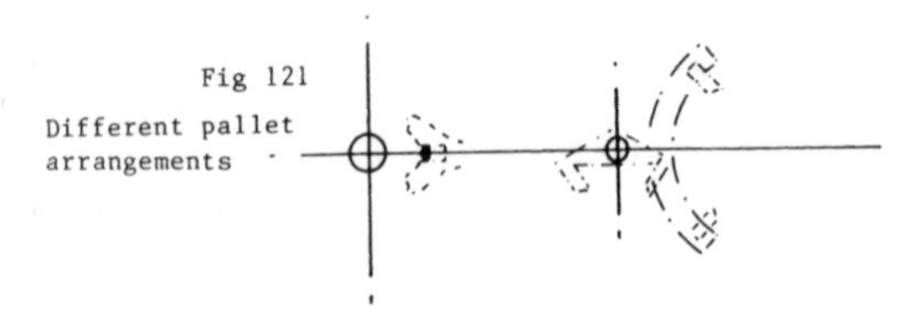

Fig 121
Different pallet arrangements

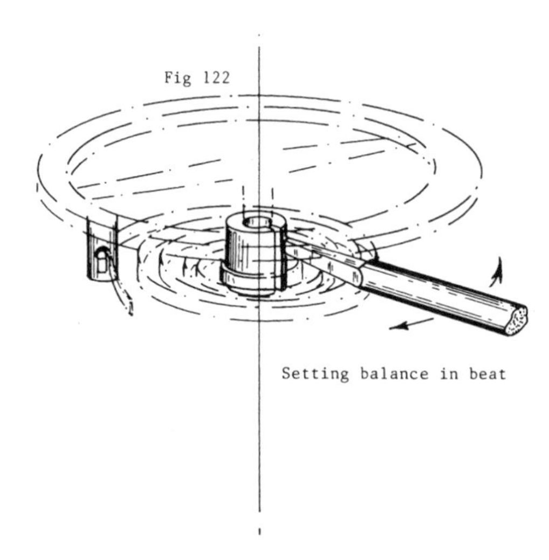

Setting balance in beat

Adjusting balance for beat

Fig 122 shows the arrangement of the balance wheel and spring on both cylinder and lever escapements. The beat of the balance can be altered by moving the collet that holds the inner end of the balance spring so that the relationship between balance and spring is changed. Relieve the balance of any impulse by passing a thin wire through the crossings of the wheel that drives the escape pinion. Make a thin wedge that can be fitted into the slot that you will see in the brass spring collet. This wedge is then used to turn the collet on the balance staff so that the spring tends to turn the balance less in one direction and more in the other. You will have to decide, by observation, in which direction the collet has to be turned to correct a balance that is out of beat. A major cause of problems in platform escapements is dirt and old oil (see Chapter 2).

THE 400-DAY CLOCK

This is a Graham dead-beat very similar to that used in the Vienna regulator; however, it rarely has reversible pallets. Fig 123 shows the positions of the major adjustments that can be made. The beat is set by turning the upper mounting of the suspension spring (in this case the screw A to left or right). The angles ß and ɣ should be evenly arranged about the 'at rest' position; however, the small bar at the top of the anchor is not necessarily upright and various other damage may have occurred. The important thing is that

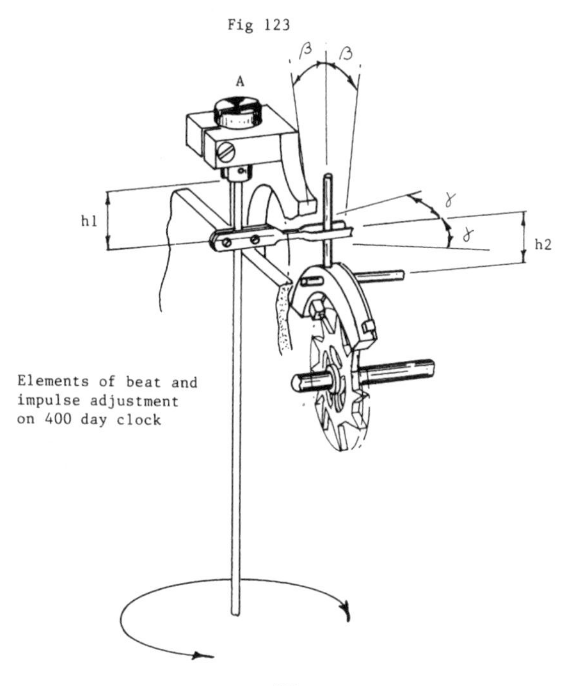

Elements of beat and impulse adjustment on 400 day clock

the rotating pendulum makes the same amount of rotation to release the entry pallet as the exit pallet. Raising the forks and increasing h^2 will decrease any tendency to skip escape teeth, it also increases the angle of rotation needed from the pendulum to release the teeth. Lowering the forks has the opposite result.

Almost all the faults that appear in a 400-day clock that are not associated with damage are found in the area of the escapement. Assuming that the clock is in beat you should watch the operation of the escapement bearing the following in mind:

1 The forks must be a loose fit, not a sloppy one.
2 Every time the pendulum rotates it lifts slightly, and the greater the angle of rotation, the greater the lift and the more the escapement has to work to keep it rotating.
3 If the forks are lifted the pallets will find it harder to move them, and the pendulum will rotate through a greater angle.
4 The reverse is true when the forks are lowered.
5 Because the fork (or crutch) is mounted on a flexible spring there will be a tendency to bounce if the pallets go over with an excess of energy, and even good unworn pallets may show a tendency to skip an extra tooth through the beat.
6 Movement of the forks up and down the spring is used to control the action of the pallets.

Although the spring that the pendulum hangs on is delicate and should not be kinked or damaged, it is not often that such damage is the cause of clock failure. It is advisable to put the clock in beat and then check the operation of the escapement and, if possible, get it working before replacing the spring – which ought to be replaced after the problem is solved.

Charles Terwilliger's book *400 Day Clock Repair Guide*, published by Horolovar, is an excellent introduction to this type of clock and contains invaluable information on styles, dates and the suspension wires. There are 23 different thicknesses of suspension in the Horolovar range, and in addition some of the more modern clocks still have wires available from their manufacturers. However you will occasionally find that none of the available sizes of suspension wire will suit a particular 400 day clock, which leaves you with the choice of modifying a standard spring by stoning it down in thickness, or modifying the mass of the pendulum. The latter is preferable (though not always easy) simply because a non-standard spring is clearly as liable to damage as a standard one and furthermore is tedious to make. Altering the mass of the pendulum so that a standard spring may be used is simpler, and as long as it makes no obvious change to the appearance of the clock, perfectly acceptable.

7
Verge Escapements

DESCRIPTION

The verge escapement consists of a crown wheel (which differs from the escape wheel of the anchor escapements by having the teeth cut from a short cylinder so that they lie parallel to the arbor and not radial to it), and the verge staff. It must not be confused with the American verge or bent-strip escapement pallets.

Fig 124 shows the crown wheel and staff. The latter is an arbor with two flags mounted on it so that they will engage the teeth of the crown wheel passing beneath. When used in an English lantern clock or a bracket clock, the verge has a short pendulum fitted directly to the staff, a knife-edge bearing supporting this end of the staff and a plain pivot supporting the other. Dutch verges and a number of other Continental types have crutches fitted so that the verge is not directly connected to the pendulum, the crown wheel may be mounted on a vertical arbor with a crutch similar to normal recoil movements or mounted on a horizontal one with staff and pendulum hanging vertically. A short crutch juts from the staff and engages the pendulum rod in a fork. Verges with long flags are also found on the Continent, which gives the pendulum a smaller arc of swing than is usually associated with the verge escapement.

Most verges are recoil escapements and not very accurate, but not all; Fig 125 shows a dead-beat verge (a rather rare escapement), which is a fairly bulky structure. Wear at the locking and impulse faces can be repaired by soldering on an L-shaped piece of gauge plate in the shaded area.

Crown wheel and arbor

Crown wheels are generally made from castings, although some (the older versions) have been fabricated from a strip of beaten brass bent into a circle and then brazed onto a disc. Very early ones were made of iron, but these are unlikely to be seen outside museums and specialist collections. The faults are similar to those found in other escape wheels – bent teeth, pitch errors, short teeth and, of course, improper repairs.

Fig 124

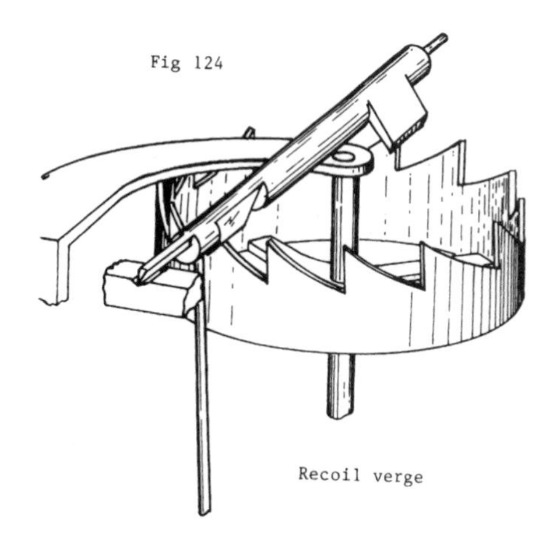

Recoil verge

Fig 125

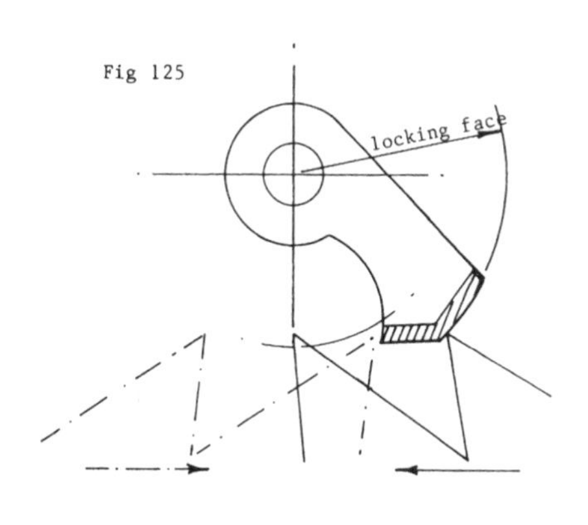

Dead-beat verge

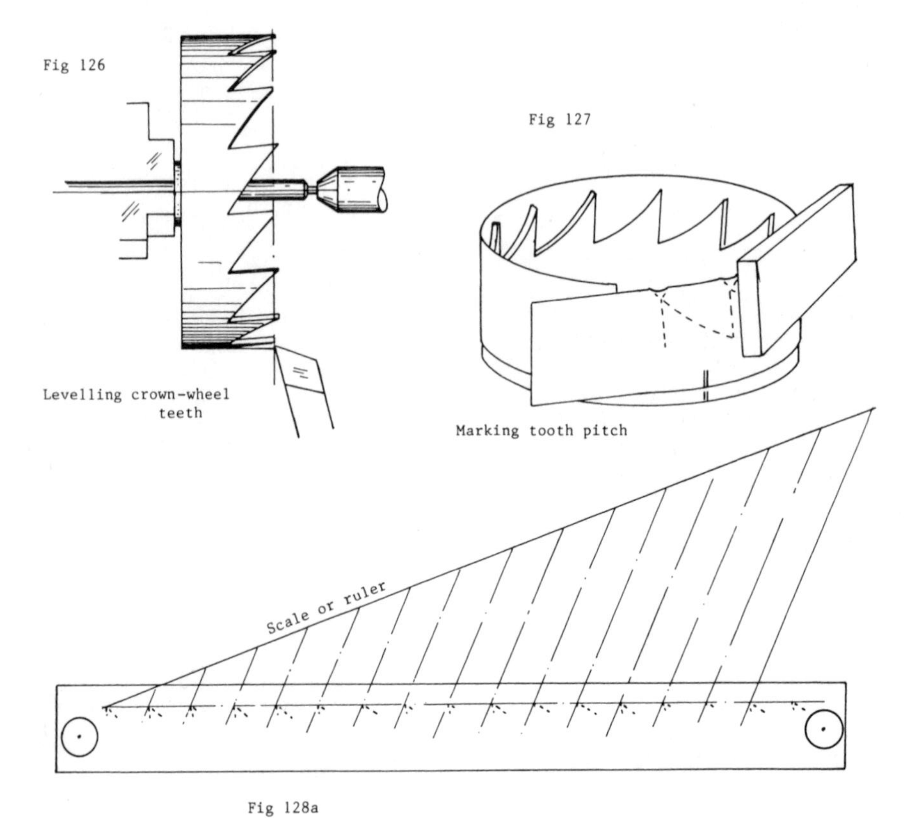

Fig 126

Levelling crown-wheel teeth

Fig 127

Marking tooth pitch

Fig 128a

Checking pitches against a scale or ruler

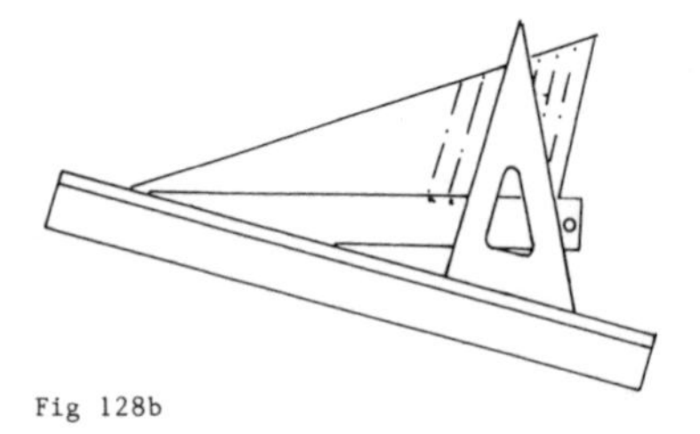

Fig 128b

Position of straight-edge,set-square and paper trace

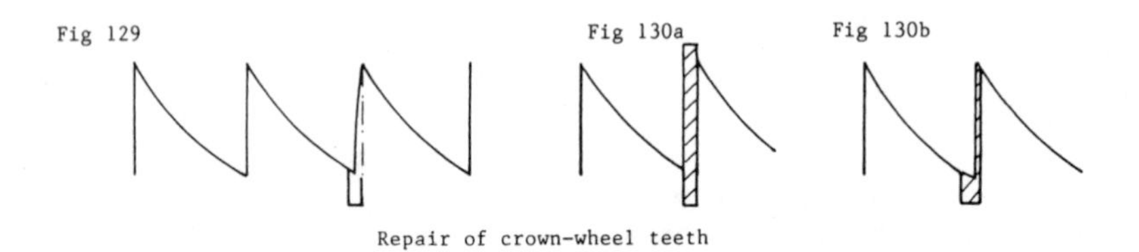

Fig 129 Fig 130a Fig 130b

Repair of crown-wheel teeth

Bent teeth can be straightened with smooth-jawed pliers until the shorter, upright side of the tooth has a flat surface again. This face is usually raked, ie it leans away from the vertical when the disc of the wheel is horizontal; unlike the anchor-recoil wheel, the raked side is the leading face. After straightening the tooth take a *very* fine file, or 400 to 500 grit emery paper backed up with a metal strip, and polish both surfaces of the tooth.

Pitch errors can be spotted by watching the action of the verge flags as the wheel turns beneath them; if the pitch is irregular, the amount of drop on each tooth will vary. Do not be too pernickety; there will have been errors built into the original cutting of the wheel teeth, the errors varying in size according to the quality of the maker and his tools. What you are looking for are gross errors and these will occur where the wheel has had new teeth inserted, where teeth are badly damaged, or where the wheel has been bent from a true circle by mistreatment or a crossing-out cracking due to age. Having found an area where the teeth are altering in pitch, check that they are all of the same height before deciding to correct; broken tooth-tips will also cause varying drop. Put the arbor in the lathe and face off the teeth until the lathe tool is just scraping the shortest (Fig 126), then use a file to remove metal from the sloping back of the teeth until they all have points again. Leave a flat on each tip of about 0.125mm (0.005in).

Now that you are certain that any uneven drop is due to pitch errors alone, look again at the area where this is found. Prepare a piece of paper with two-sided adhesive tape, and cut it into a strip that can be wrapped around the crown wheel so that it covers the wheel from base to tooth-tip. Mark the position of each tooth-tip on the paper by pressing a smooth piece of metal onto it, crushing the edge of the paper between the metal and the tip (Fig 127). Use a pencil to put location marks on paper and wheel so that you can find the right position again. Now unwrap the paper from the wheel.

You now have a flat trace of the pitches of the teeth. A little simple geometry will show the correct pitch for each tooth, but you must decide where the pitch begins to go wrong and when it is right again. Use the two-sided tape to stick the paper down onto a sheet of cartridge paper, then all that is needed is a ruler, a sharp pencil and the instructions in Figs 128a and b. If you remember your geometry from school days, you will recognise the old method of dividing any line into equal parts.

Select for correction those teeth whose pitch errors exceed 10 per cent of the pitch, number the teeth on the wheel and on the paper strip for accurate identification, and then use a hacksaw to produce the shallow saw-cut shown at the root of the tooth (Fig 129). One edge of this cut will be in line with the front face of the tooth. Lightly clean the front edge of the tooth with a file or abrasive, beat a piece of 70/30 brass until it is thin enough to enter the slot, but still thick enough to provide the required new front face of the tooth. Clean it, and tin it with soft solder. All that is now necessary is to insert the new piece of brass trimmed to be a little larger than needed in height and width, and to warm it with a gentle gas flame until the solder makes a junction with the old tooth (Fig 130a). When the solder is cold, remove the spare brass and file the front of the new tooth face with a smooth file, using the paper to obtain the correct position of the new tip. (Fig 130b). Repeat the operation on all inaccurate teeth, one at a time. It sounds a slow process, but three or four teeth can be replaced in this way in less than an hour.

VERGE FLAGS

Repairing flags

The flags are the impulse faces and, as we have seen, they engage opposite sides of the crown wheel and receive the impulse alternately. Verges made before the middle of the eighteenth century have a wide angle between the two impulse faces of around 110 degrees; more recent verges often have a smaller angle of between 60 and 80 degrees. You will have to decide from the remnants of the original flags, what angle was used by the maker. If the damage is so great that there is no reliable evidence on the flags, make the angle conform with the apparent date of the clock.

Wear on the flags can often be repaired by simply softening the steel, filing back the flag, and soldering on a flat piece of spring steel using the same method as for an anchor escapement (Chapter 5). The pieces can be made as wedges if the edges of the flags are very badly damaged (Figs 131a and b).

Replacing the verge

If the flags are very badly damaged, it is very probable that the pivot and knife-edge are in poor condition too, and the simplest thing to do will be to make a new verge to the old pattern. Three methods follow:

FLAT STRIP

This is probably the oldest method of making a verge. Take a piece of gauge plate or flat-ground stock and cut it to the pattern shown in Fig 132. Note that it must be long enough to machine the pivot at one end and file the knife-edge at the other. Put one end of the steel in the guarded jaws of a vice, heat the strip mid-way between the flags until dull red, and then twist the other end until whatever angle you have decided to make the flags is opened up between them (Fig 132). The staff or arbor can be turned along its length, or filed into round; the pivot end must be turned (Fig 133). The diameter of the staff is of little importance, it must be left strong enough not to bend in use and small enough not to interfere with the crown wheel. Making the knife-edge is dealt with below under a separate heading.

BUILT-UP FLAGS

These are a substitute for verges that were forged originally, but are not, of course, limited to that. Take a piece of silver steel and turn the pivot, making sure that the stock diameter is at least twice the proposed depth of the knife-edge which is usually 1.5mm to 2mm (0.06 to 0.08in) deep. The flags can then be cut from gauge plate and fastened to the staff in their correct positions. However, there are different ways of doing this:

1 Choose steel plate that is as thick as the staff and hacksaw a groove into the face of the two flags that you cut from it (Fig 134), using two or three blades side by side in the hacksaw frame. The groove is positioned so that the staff will sit in it and the flag will be long enough to take impulse from the crown wheel. Silver solder the flags to the staff and then file the impulse face as shown in Fig 134. Note that this shows the impulse face passing through the centre-line of the staff, which is the normal method.
2 The plates used for this method are thicker than the staff so that they can be drilled and slid onto it. The impulse face does not pass through the centre-line of the staff (Fig 135).
3 One piece of steel is used and turned eccentrically to leave short cylinders of the original diameter from which the flags can be filed.

Method 1 produces an arc of pendulum that is equal (considering the working arc only) to the impulse angle. It is not easy to harden the flags because of the likelihood of melting the silver solder in the process. Hold each flag in turn, at the junction with the staff with a pair of old pliers (Fig 137), then heat its working end to red heat and quench in water (for speed). The mass of the pliers should prevent the joint from melting as the hardening proceeds, subsequent tempering to amber will not affect the joint at all.

Method 2 produces an arc of pendulum that is less than the impulse angle (shorter arcs are preferable for good timekeeping). It also enables you to harden and temper the flags and then fasten them to the staff with soft solder or an industrial adhesive. Soldering automatically tempers the flags blue.

Method 3 can be used to give an angle that passes through the centre-line or not, as you please, and hardening is no problem at all; but the method does call for more expertise in machining than the other two. Tempering is to amber colour.

PIVOTS AND KNIFE-EDGES

Pivots are treated in the same way as detailed in Chapter 4. Knife-edges are formed from the staff and simply consist of a short triangular-sectioned extension to it, the lower edge of which must be straight, without either chipping or burring (Fig 138). It should not be absolutely sharp – a radius of a hair's breadth along this edge will prevent excessive wear on the bed (the bearing that it rests on) and not affect the performance of the verge to a noticeable degree. Occasionally these extensions are broken off as a result of the maddening sort of accident that happens when someone is looking over your shoulder – or perhaps it was just sheer clumsiness.

Make up a new knife-edge from a short

Fig 131a

Fig 131b

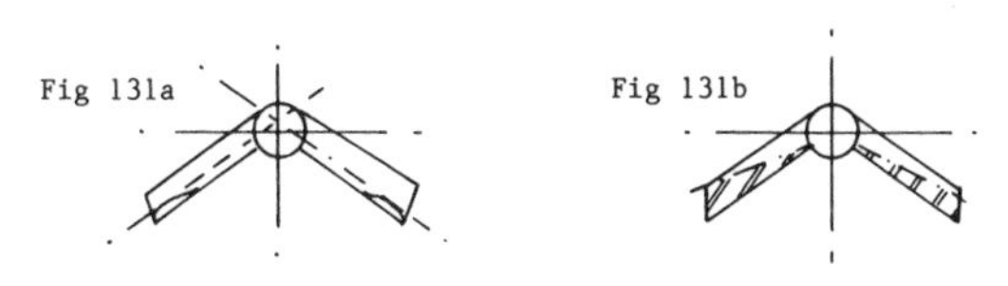

Alternative preparation for soldered facings

Fig 132

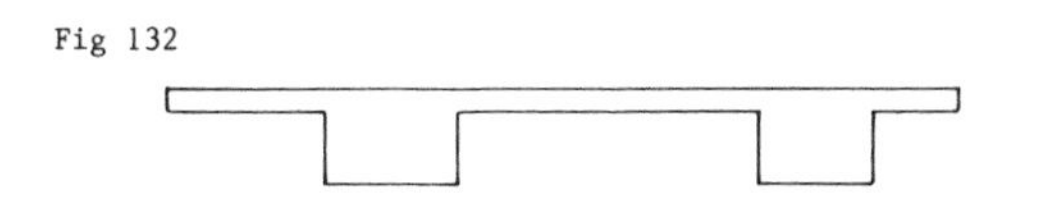

Verge staff and flags from flat strip

Twist

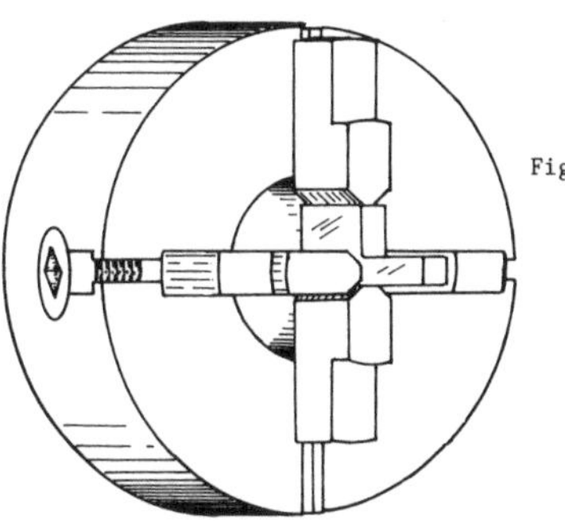

Fig 133

Holding flat strip to turn pivot

Fig 134

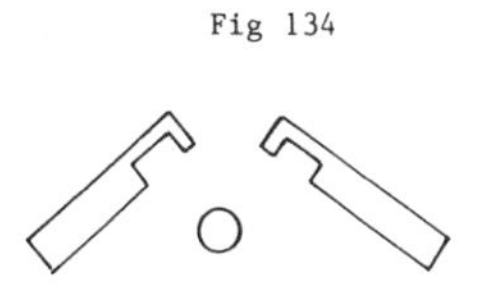

One method of making flags

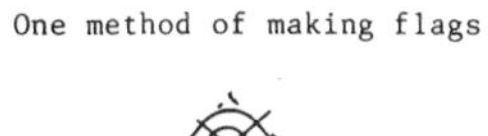

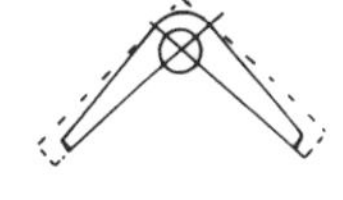

Fig 135

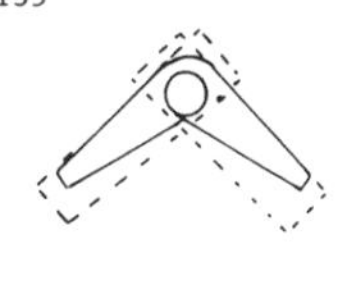

Fig 136

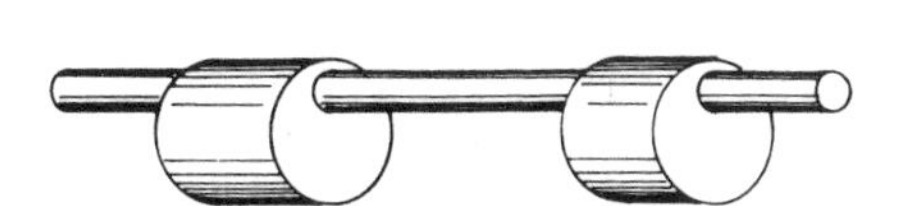

Two more methods of making verge flags

Fig 137

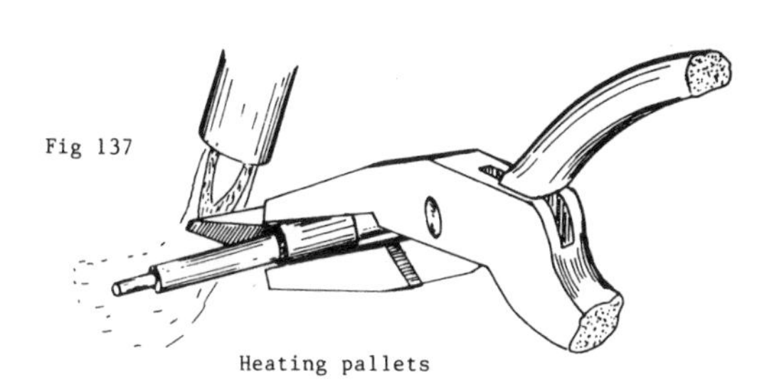

Heating pallets

Fig 138

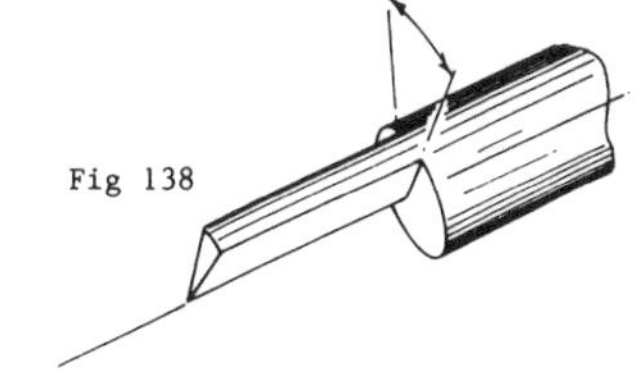

Knife-edge bearing

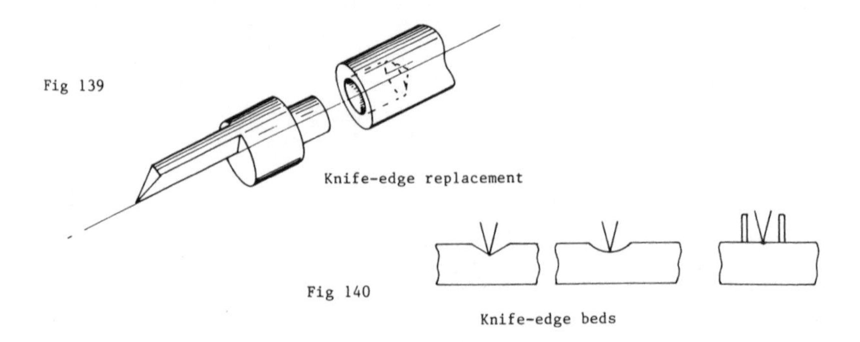
Fig 139

Knife-edge replacement

Fig 140

Knife-edge beds

piece of silver steel of the same diameter as the staff; at the other end turn a diameter that can be plugged into a hole to be drilled in the old staff. The plug end should be about twice as long as the diameter (Fig 139). The original staff will have to be softened before it can be drilled, and it will also need to be shortened so that when the extension is fitted to it the distance from pivot shoulder to knife-edge shoulder will remain unchanged. Harden the knife-edge and temper to a very faint amber; it is fastened in position with an industrial adhesive. Please note the angle on the end of the knife-edge; this contacts an end-plate or apron on the clock and holds the verge longitudinally without excessive friction.

BED AND APRON

The plate that prevents the verge sliding out of the movement is called the apron, and it is usually made of brass and highly decorative. The apron rarely needs anything but a light stoning on the face that makes contact with the end of the knife-edge. Its squareness is not important, and it only needs a polish where the knife-edge touches.

The support for the knife, the actual bearing (bed) takes a lot of wear. Many of the beds fitted to old bracket clocks are of brass, although the reason for this is uncertain. You will also find that the maker very rarely bore in mind the possibility of needing to renew the bed, and several repairers will probably already have dovetailed in new brass pieces. The kindest thing a repairer can do for an old bed that has been treated to several replacements is to select the best looking of the dovetails, remove the old bed, and make a new one to fit that dovetail.

If a clock is not moved, the knife-edge will stay in position on a perfectly flat bed; but clocks are moved, and it is impossible to leave a clock without some means of positioning the knife-edge. I prefer a very shallow vee or a radius. Banking pins are sometimes used, but if the knife is bumped it will come up against the pin and the staff will not always move back of its own accord (Fig 140).

Vee or radius, the knife-edge must be in contact across the full width of the bed. If the clock was designed with a self-levelling bed, there will be no problem, but this is a rarity. After making the bed and adjusting the height to suit the passage of the crown wheel beneath the flags, lightly scrape the knife from side to side in the bed. If it does not scrape evenly across the full width, take a piece of emery paper backed with a round rod or a strip of brass and remove metal at the places that the knife touched. Do this until the bed is in full contact with the length of the knife. If this is not done the knife will cut its own level, but in the form of a notch that will bear on the sides of the knife. If a steel bed was fitted in the first place, then use dead-hard steel.

LOWER BEARING FOR CROWN WHEEL

The top of the arbor carrying the crown wheel is held with a cock and a normal pivot hole. The bottom has a hard steel pad beneath the end of the pivot to take the weight of the crown, and the height of this can usually be adjusted by means of a wedge or a screw. This should be up to the top of its adjustment when a new bed is fitted so that, as the bed is dressed over the years, the dropping of the verge staff can be taken up by lowering the crown wheel (Fig 141).

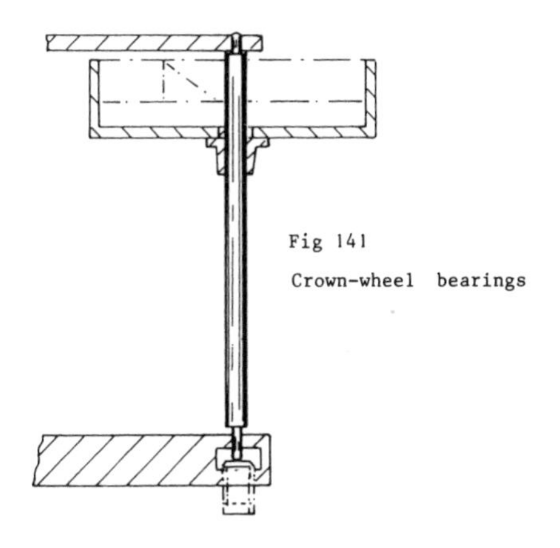

Fig 141

Crown-wheel bearings

DRIVING POWER

A verge that is fitted with a knife-edge will tend to lift when operating. If the weight or spring drive is excessive, this lifting will cause all sorts of problems. Reduce the weights to whatever will work the clock reliably. Springs are more difficult, inspect them for signs of a replacement and, if it does appear that a new spring has been fitted at some time, refit with a lighter one. If, however, it is obviously the original, you will have to leave it as it is. Indeed if the weights in a clock are in any way special and not the normal anonymous lumps of lead, they must be left alone as being part of the history of the clock.

8
Train Faults and Gears

Clocks have a limited amount of energy available to them, there is very little to spare for any increase in frictional losses in the train. But the problem of a train that has so much friction that it no longer serves to drive the escapement can be solved in two ways.

The first way is to add more power, either by weight or by changing the spring. No matter how much trouble is involved do *not* make a clock work by increasing the power above what was there before. In fact, because we no longer have to cope with oils that gel inside two or three years, many old longcase clocks can have the load on their movements reduced as long as this does not affect the appearance in any way. The original maker knew that he had to provide a surplus to allow for the failure of the lubricant.

The second solution is to examine the clock carefully and then strip it and carry out the corrections needed.

MOTION WORK, DATE AND MONTH WORK

Examination for faults

Deal with the 'outside' gearing first, the motion work and any date or month work (Fig 142a, b, c, d). Any date or month work driving an indicator or dial that is supported on the dial will have to be tested with dial in place. Those clocks that have a leaf-spring drive to the minute pipe should have this removed so that the hands turn quite freely (Fig 142b), other types can be made a little more free by lubricating the centre arbor inside the minute pipe. The aim is to make sure that any 'lumpiness' is derived from the wheels under test, and not from the movement. Fit the minute hand and then gently turn it to operate the motion work and the indicators on the dial. You will need to take it through at least twenty-five revolutions of the dial, applying the turning force as close to the centre as possible both to protect the fairly delicate minute hand and to make the most of any resistance that shows up.

Failure of the date or month work (Figs 143, 144) will very likely be due to the dial changing its position relative to the movement, or sometimes the pin or finger that engages on the teeth of the dials or the star wheel. Check the dial feet and the posts that carry levers and wheels associated with the indicators. Remember that what we are testing for is a failure that is stopping one or other of the clock trains; simple failure of the indicators will be a matter of missing pins, dropped dial or missing wheels.

Motion work that stops the clock can do so as a result of a broken or bent tooth, or butting caused by the wheels' moving apart on a worn or bent bearing. Trouble due to broken teeth is not always obvious, because in the motion work they are unlikely to have been strained mechanically, the fault is more often a result of corrosion. Brass is prone to corrosion by ammonia if the metal is hard but not stress relieved, the failure being along the boundaries of the grain. This leaves a very close jig-saw fit of the two parts and, if the fault has not quite parted one piece from the other, the tooth will flap open on the remaining metal and close up so well that the break is not visible without a magnifying-glass.

Movement trains

Remove the motion work, escapement and front-work from strike or chime; if the clock is spring driven, remove the ratchet wheels, then place the movement on the bench, supporting it if necessary so that no part of the train fouls the bench or support and there is no likelihood of it falling over. Apply as light a load as possible to the teeth of the great

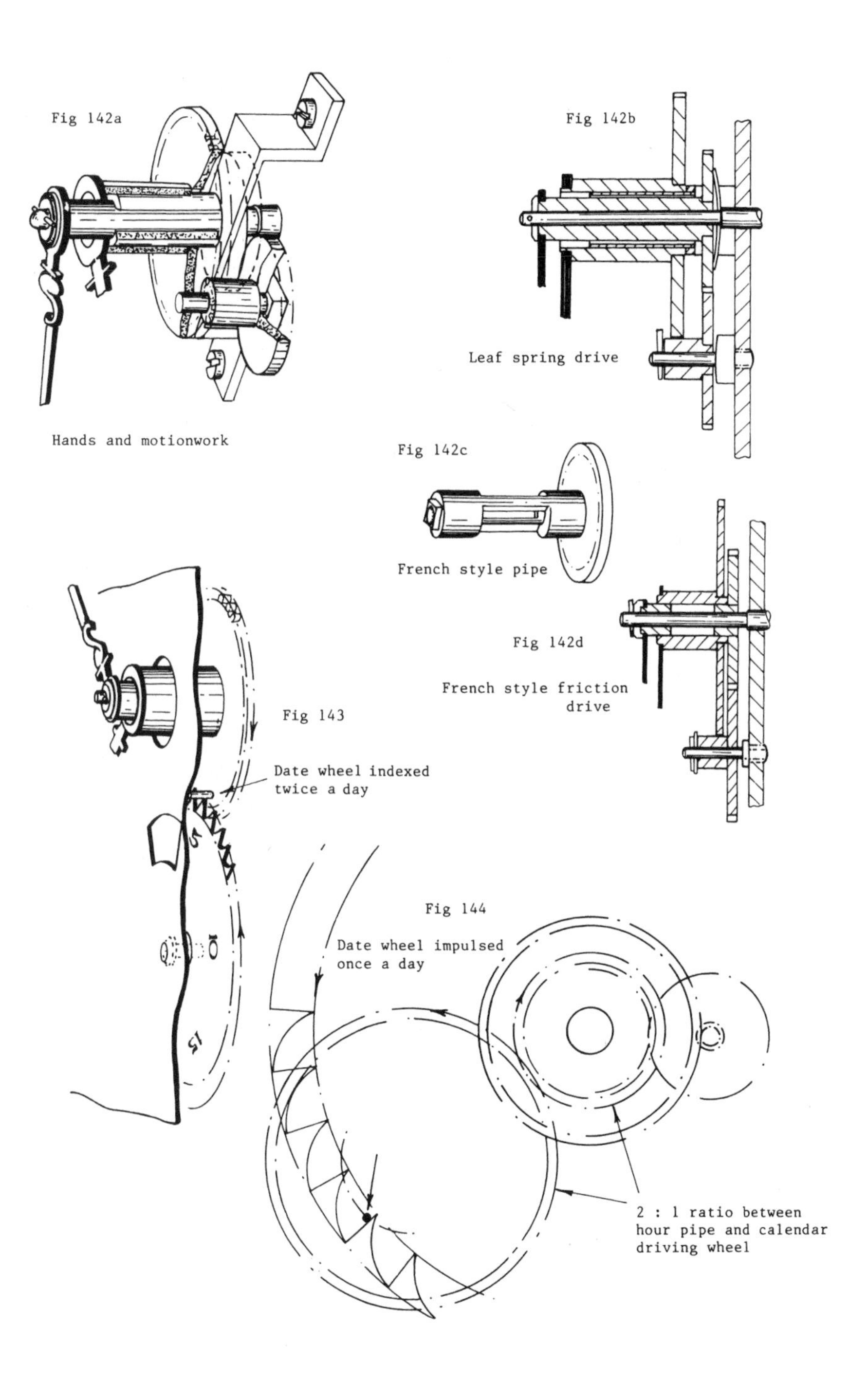
Fig 142a
Hands and motionwork
Fig 142b
Leaf spring drive
Fig 142c
French style pipe
Fig 142d
French style friction drive
Fig 143
Date wheel indexed twice a day
5
10
15
Fig 144
Date wheel impulsed once a day
2 : 1 ratio between hour pipe and calendar driving wheel

wheel so that the train only just turns. In the case of a weight-driven clock it is simple enough to hang a small weight from the barrel and increase it until the gears rotate. However, if you take a thin piece of pivot steel or a stiff wire, hold it 75mm or 100mm (3in to 4in) from the end and press down on the teeth of the great wheel or barrel wheel, you will have very close control over the force applied. The point of the test is to discover whether the amount of force needed for rotation varies much during the course of a complete great-wheel revolution. Holding a stiff wire by the end and using that to apply pressure you must work quite hard, and any differences will be immediately obvious. There will be variations as a result of the fluctuations of force transmitted from tooth to tooth, but these should be minor.

It is very difficult to define acceptable fluctuation, common sense must be exercised; but a movement that demands twice the driving force at one point is not going to have a particularly stable escapement (unless it has gravity escapement), and the time-keeping will vary. On the striking or chiming train the effect will be slow striking sometimes and fast striking at others. A little thought will indicate what is acceptable in the way of difference in striking speeds. In practice, this amounts to nothing that is discernible when separated by a quarter of an hour.

Three types of movement are possible – relatively smooth, variable and partial – the latter ending in a complete stop and failure to start within the capabilities of the clock's driving mechanism. In other words, this means a weight-driven clock that will not restart after the train is stopped and the normal running weight is applied, or a spring-driven clock that behaves in the same fashion in the last full turn of the great wheel. The use of restart is important because you must not make use of the momentum of the rotating train for a useful test; the going train comes to a stop at every beat and the strike. At the end of every striking period, the clock must be capable of starting again.

Smooth movement of train

Almost certainly the teeth, pivots and centre distances of the train gears will be in reasonable condition. Examine the pivots for wear as laid out in Chapters 3 and 4 and then thoroughly clean the train and the plates, pegging out the holes afterwards and burnishing the pivots. As long as no dirt creeps in whilst assembling the clock, you can discount the train as being the cause of any future troubles. For one final check, hold the last wheel in the train still and apply a little more than normal running load to the great wheel. This should show up anything unusual such as slipping wheel or collet, or a loose redrilled pivot.

Variable movement of train

It is best to remove all the wheels and pinions of the suspect train and try them separately in their pivots. If the latter are in good condition, they will not be badly grooved, the holes will not be oval, the arbor will fall from side to side with an audible click as you turn the plates on their side and, when the wheel is spun in its bearings, it will come to a slow stop, not a sudden one.

Supposing that all arbors with their wheels and pinions pass this test, the actual meshing of wheels and pinions for each pair must be examined. Start with the great wheel and whichever arbor – centre or intermediate – is next to it. Put both in their correct positions between the plates and gently turn the great wheel, using a very gentle touch so that you can feel any roughness in the mesh at any point. If nothing shows, turn the great wheel fast and rest a piece of pivot steel or a pegging stick on the other arbor to apply a light braking load. There should be no bumping of the teeth as they mesh and when you remove the brake and the pressure from the great wheel, the pair should come to a gradual stop. The reason for the brake is that if the driven wheel is left completely unloaded it will not behave as it does in the complete train and may also 'catch up' on the driver as your turning of the wheels alters; in this case any butting that occurs would be irrelevant, because the clock cannot repeat this when operating as a clock.

Go through the train, trying each pair in turn and only having two arbors in the plates at a time.

This test will reveal any off-centre wheels or pinions, any bent pivots or eccentric pivots, any bent teeth, excessive wear or pivot holes or teeth and, of course, any faulty workmanship of a previous repair.

Complete stop

If you can maintain the driving pressure on the train, it will now be possible to find out just how far through the train the driving energy is reaching. Make a very small movement of the last wheel in the train, in the direction in which it should be going. It is sufficient to move it far enough for the teeth of the next wheel down to have space between them and the pinion leaves; if a wheel can move away from the teeth of the next wheel down when driving load is being applied, then clearly no energy is reaching it (Fig 145). This wheel can be passed and the same test made on the next wheel down. As soon as you move a wheel and the next one down moves after it, you have located the trouble. It may not be the fault, but at least it is a starting point.

A tooth fault (bent tooth, butting) will reveal itself as the driving wheel suddenly comes free when the next arbor up is turned. Mark the position of the wheel when stopped by the failure – a pencil is best, not a scriber. If the stoppage is repeated as each tooth comes into contact with the pinion the problem is butting, very probably caused by worn pivots or a badly placed bush.

An eccentric pivot will give a tightness on one side of the wheel and none at all on the other, so will a bent pivot. Pivot holes that are not exactly opposite each other will give the same result.

If a wheel is running true about its centre, but waving from side to side, it will only give trouble if the face touches some other part. Indeed there is an advantage in a wheel that runs in this way, in that the wear is spread over a longer portion of the pinion leaves (Fig 146). However, the opposite is true of the contrate wheel, ie the wheel that gives a sideways drive to the escape pinion of a platform escapement and has teeth cut from a short cylinder, not a disc. In this case the truth of the teeth to a plane at 90 degrees to the arbor is the important factor; eccentricity simply spreads the wear along the length of the pinion (Fig 147).

Chapter 4 gives details of pivots and shoulders and the friction loads developed there. In a normal longcase clock, these can be more than half the clock's total frictional loss.

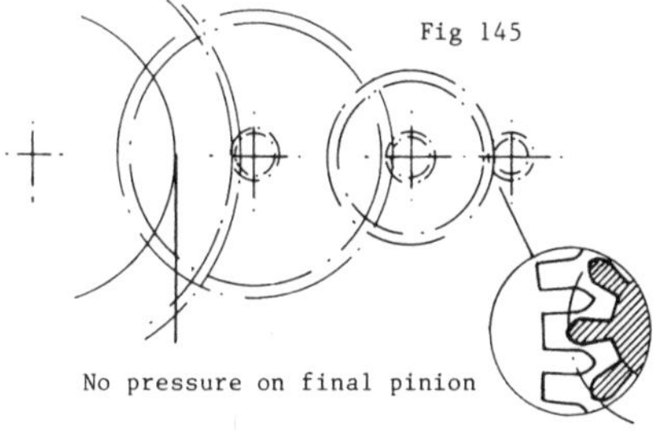

Fig 145

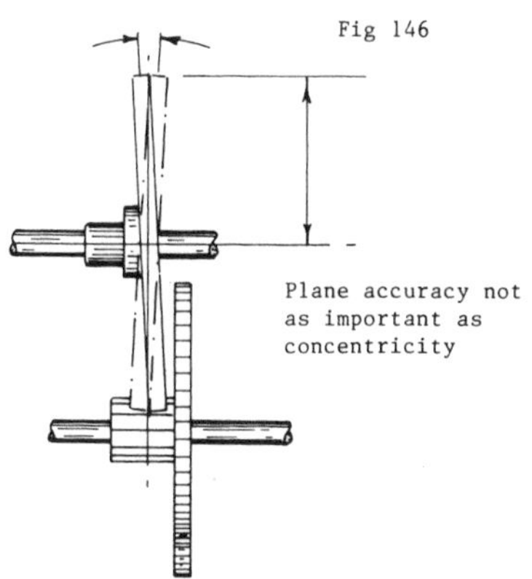

Fig 146

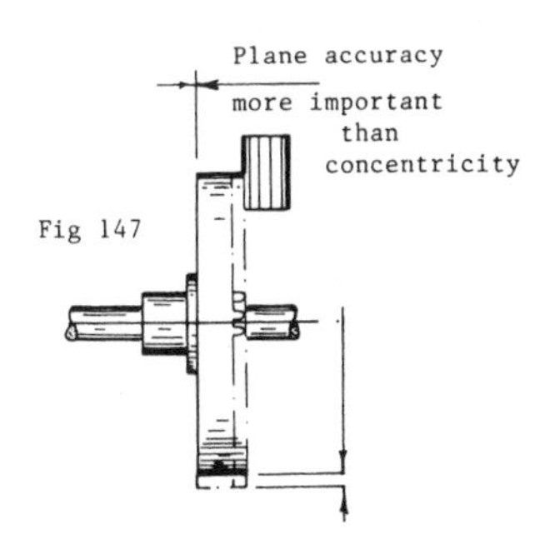

Fig 147

CORRECTING ECCENTRICITY

Wheels and pinions

Many old British-style clocks have a wheel mounted directly on the pinion. If there is any reason to remove this wheel, for instance so that it can be softened for re-pivoting, it often will not go back and run true. This is because the pinion teeth have cut into the bore of the wheel and in doing so have raised a burr on the inside, or in some fashion cut into the diameter, of the hole. The original wheel was very probably turned after wheel mounting and then cut, as was common practice on French round movements; now the remains of the bore bears no true relationship to the outside of the wheel. Set the wheel in a piece of scrap brass that is held in the lathe chuck and has been bored to fit the outside of the gear. Fix in place with shellac or hard wax by warming the wax and then letting it cool while the wheel is held in place. Now bore as little out of the hole as will make it a complete and true diameter (Fig 148).

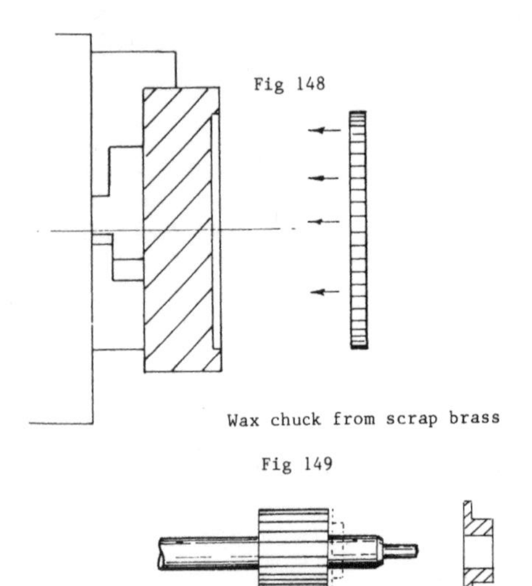

Fig 148

Wax chuck from scrap brass

Fig 149

Pinion faced back and collet made to mount wheel upon

The wheel now has a bore that is true to the outside, but it is too big to fit back onto the pinion. The only reasonable solution is to make a short bush that can be pressed onto the pinion and locked there, and turn it until the wheel will slip on (Fig 149). To ensure truth during this turning operation hold the arbor – if it is concentric to the pinion – in a collet chuck, or hold the far end of the arbor in a three-jaw chuck and support the pinion pivot in a hollow centre held by the tailstock. The wheel can be locked securely with Loctite or some other cyano-acrylic adhesive.

A pinion that is cut from the same piece as the arbor, can only be made concentric to the pivot by drilling and re-pivoting. Since the only way in which it could have become eccentric after original manufacture is by bad pivoting or bending, this solution seems logical enough. As indicated in Chapter 4, the only important relationships so far as concentricity is concerned are pivot to pinion, pinion to wheel or wheel to pivot; the arbor is only a piece of material that relates one part to another, it does not affect the accuracy of that relationship.

Plates and pillars

As previously remarked, holes that are not in line with their complementary bearing in the other plate, will give much the same results as eccentric or bent pivots. The fault may come about as an effect of bad bushing, or of a fairly traumatic accident to the whole movement that has shifted the plates so that they are no longer in line with one another. Such an accident may bend pillars, the area of plate surrounding the pillar mounting, or tear the pillar out altogether. The correction of pivot holes is covered in Chapter 3 and the treatment of damaged plates and pillars in Chapter 1.

DAMAGED WHEELS AND PINIONS

We have dealt with eccentric wheels and pinions, but not with gears that have damaged or missing teeth. Bent teeth may be straightened with a screwdriver as described in Chapter 1 (Fig 11). However, it is as well to inspect these teeth very carefully with a strong magnifying-glass afterwards. If they have cracked, the crack will almost certainly continue beyond the visible extent, and during use will grow until the tooth drops right off. Pinion leaves hardly ever bend, they snap right off, which, at least saves any heart-searching over the condition of the gear. Pinions, however, will crack part way along the length of a leaf, giving very poor drive to the train when used. Any pinion that, from tests on the train, is seen to be part of a suspect pair (wheel and pinion), should be examined under a magnifying-glass in case the trouble is a cracked leaf. These are often very difficult to find.

I can see no virtue in silver soldering new teeth onto old pinions. The repair is inadequate, the temper poor, and it is simply laying in trouble for another repairer. Large pinions from tower clocks can have the metal built up by metal spraying or – if you are certain of the quality of the steel – welding. The new leaf is then filed from the newly deposited metal and the whole heat-treated as usual. Case-hardened wrought iron is, in my opinion, quite unsuitable for welding and so are most steels of the eighteenth century and earlier. Metal spraying retains the original state of the metal if the heat is not allowed to become too great at each pass.

Small pinions, such as those in domestic clocks, will need replacement if they are broken. Study the form of the other pinions; most of those made after 1770 will have been cut with a form tool and perfectly acceptable replacements may be obtained from a number of wheel and pinion makers. Any clock prior to that date presents you with the decision of preserving the original form of the pinion and making the replacement by hand, or compromising the design and using a machine-cut pinion. The choice for the more common clock is not difficult, many of them have been compromised already!

Hand-cut pinions

Making hand-cut pinions is not a very difficult task. Turn an arbor and the diameter required at the pinion, from silver steel. Heat it until it is dark blue and then, using a dividing head – or wrapping a tape measure around a chuck and doing a little arithmetic – scribe lines along the cylinder that the pinion is to be cut from, so that you have a guide for the top of each leaf. Make a saw-cut radially, mid-way between each scribed line. The pinion blank can be held as shown in between vice jaws. Take the saw-cut down to the root of the leaf (Figs 150 to 152).

Put the blank to one side for a moment, and make a small gauge the width of the original leaves. A piece of flat brass will do, mark a line at right angles to one face and then saw-cut down the line about half the depth of the old leaves. Use a flat file to open this cut to the width of the leaf, and maintain an even balance of the slot about the scribed line (Fig 153). Make sure that the sides of the gauge do not interfere with adjacent leaves. Now use a flat file to widen the space between the roughly cut leaves until each one matches the gauge (Fig 154). When this is done, use the same file to widen the space between leaves at the root, (the bottom of the leaf) using the old pinion as a pattern, and extending the sides of the space radially to meet the gauged width. This point should be marked on the gauge from the old pinion; it is the junction of the straight side and the curved top of the tooth (Fig 155).

Finally the top must be rounded over. There is not much point to making a gauge for this radius; if you are skilled enough to file such a small radius accurately to a gauge, you will find it easier to match the curve by eye to the old pinion, one side of which is unworn as wear only takes place on one side. To make it a little easier, file a 45 degree chamfer on each leaf (Fig 156). It should be just large enough to touch the intended curve and, if care is taken with the first chamfer, comparing it with the original pinion, it will only be necessary to make all the others the same width. Rounding the chamfer into the curve is a matter of removing metal from the angles that are left. Dull the steel by wiping your thumb on it, and then make sure that the file does not touch the three established points, the line at the top of the leaf, the middle of the chamfer, and the pitch circle. Finish by polishing with emery paper backed up with a thin strip of metal.

Repairing wheel teeth

Wheel teeth give more chance of repair than pinion leaves, unless the rims are so narrow that there is no support for the replacement. As a rule of thumb, the time taken to replace six teeth is much the same as that of making a new wheel, not including the cost of crossing-out. The decision whether to insert new teeth or make a new wheel will depend on your equipment, your assessment of the time required by the two techniques but, more importantly, on the desirability of retaining the original as against the provision of a wheel that is unlikely to lose any more teeth. Most teeth bend or break as a result of weakness caused by age and corrosion; the immediate cause may have been the shock of a broken spring or weight cord, but probably this was only the final blow.

Preserving the integrity of the movement is well worthwhile and should be applauded, but most of the clocks that circulate outside museums and major collections are working clocks and, as long as replacement parts are to the same design as the original, reliable operation is the main consideration.

Having made the decision to repair the teeth, the procedure I prefer is:

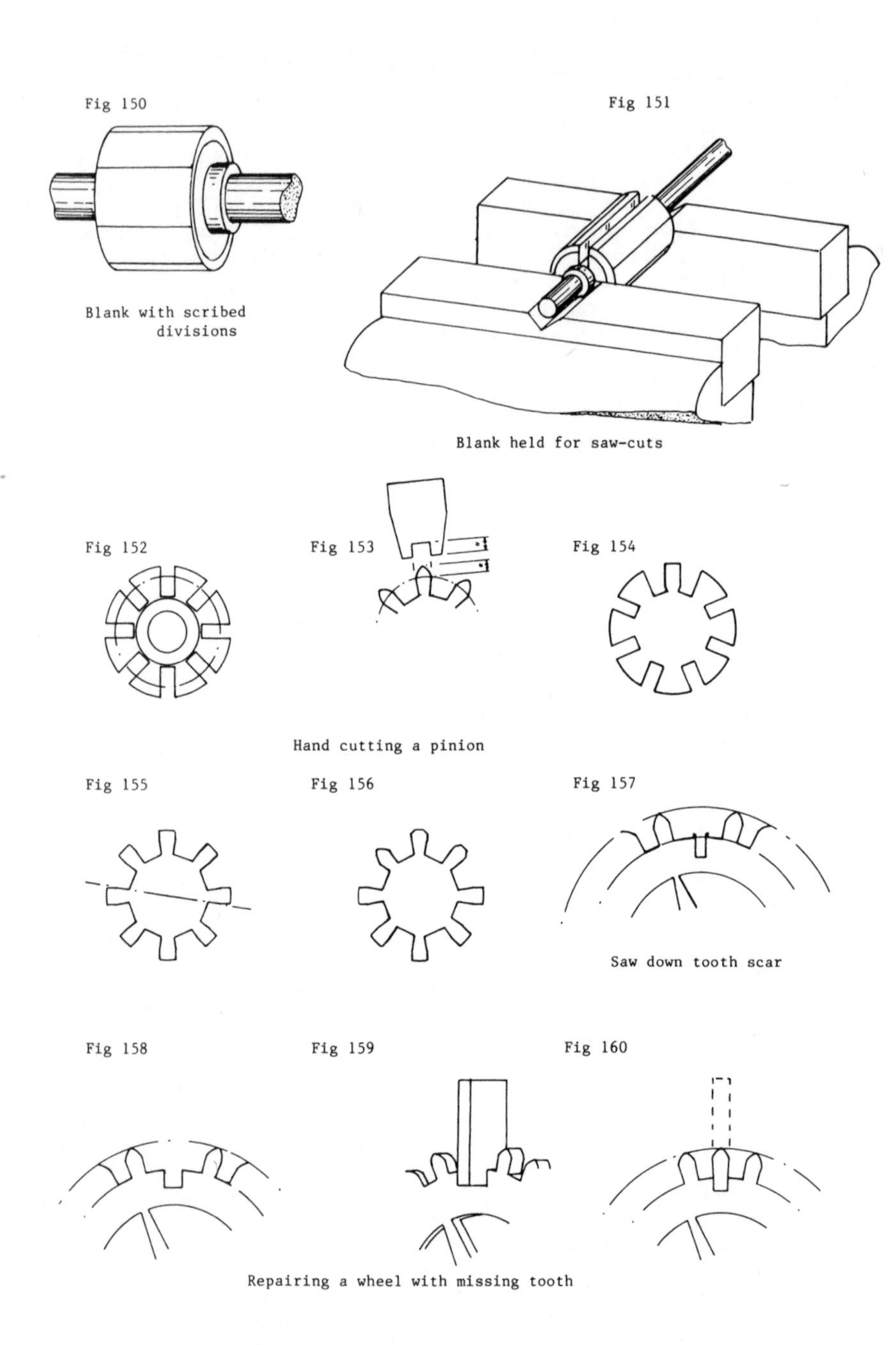
Fig 150
Fig 151
Blank with scribed divisions
Blank held for saw-cuts
Fig 152
Fig 153
Fig 154
Hand cutting a pinion
Fig 155
Fig 156
Fig 157
Saw down tooth scar
Fig 158
Fig 159
Fig 160
Repairing a wheel with missing tooth

1. Select a saw blade that is thinner than the teeth of the wheel and make a saw-cut down the radial that lies on the centre-line of the broken tooth. There is always a stub or scar that allows accurate positioning of this saw-cut. Cut no deeper than half the thickness of the wheel rim; a slot that is as deep as the width of the tooth will be quite strong enough to support the tooth (Fig 157).
2. Widen the cut with files until there is a parallel-sided slot that exactly occupies the scar width. I use a succession of knife-edge and flat files that have had one side ground smooth to obtain a good, flat-bottomed slot (Fig 158).
3. After the slot is finished take two or three strokes of the file on each side to provide clearance for the solder that will fill the joint. This will widen the slot by about 0.08mm to 0.12mm (0.003in to 0.005in).
4. Measure the width of the remaining teeth with a vernier calliper.
5. Take a piece of 70/30 brass sheet that is about 25 per cent thicker than this dimension and beat it with a planishing hammer, on a smooth steel block, until it has the same width as the teeth.
6. Reduce the size of the brass so that it sticks out each side of the wheel when put into the tooth slot and is 12.5mm to 25mm (0.5in to 1in) long (Fig 159).
7. Tin the brass with soft solder and, whilst still molten, wipe with a cotton cloth to leave a very thin film of solder.
8. Insert the brass, and lay wheel and insert on a fire-brick – or an ordinary brick that is completely dry having been kept in a dry atmosphere for several days at least – raising the wheel slightly with a little piece of scrap metal so that the insert protrudes on the top and the bottom surface of the wheel.
9. Apply soldering flux to the joint and gently warm with a gas flame or a soldering iron, being careful not to disturb the radial alignment of the replacement tooth.
10. Keep heating until the solder melts; it will become shiny. If necessary add a little more by touching a piece of solder, that has been beaten very thin, to the joint. As soon as a fillet of solder is apparent along the length of the joint remove heat and the extra solder. You do not want solder to flow into adjacent tooth spaces.
11. Let the work cool and then clean off and neutralise the flux.

This is an easier procedure than dovetailing, and a neater one. In addition it locates the teeth more accurately than can be done with dovetailing a block of metal, marking out and cutting. The use of a gauge for the tooth space, such as is often recommended, is only useful if the original teeth were divided to modern accuracies. Variations of 0.05mm and 0.08mm (0.002in and 0.003in) are common and become very obvious when more than two teeth have to be replaced.

Finish the job by cutting the over-long brass down until the new tooth is the same height as its immediate neighbours. It is quite useful to turn the last little bit in the lathe. Form the top of the tooth with files; a barette (a triangular-section file of which only the longer side of the triangle carries teeth) is very useful; polish with emery paper. The last operation is to file off the brass that sticks out of each side of the wheel so that the new tooth is no thicker than the others, and then polish on a board and emery paper. It should be very difficult to find the replacement tooth, particularly if the sides are polished clean of the solder that made the joint.

Machine-cut gears

This is not really the place to discuss the machinery needed for gear-cutting; in any case unless it is the intention of the reader to build his or her own dividing head and gear-cutter, all the pertinent information will be set out in the literature supplied with the machine. However, the calculations and methods that are used to determine the dimensions of the wheel and pinion blanks are useful, and most textbooks lack detailed information on the making of fly-cutters for wheel cutting. Cutters for pinions are almost invariably multi-tooth and should be bought from a supplier; home-made multi-tooth cutters are not terribly successful and they do take considerable time to make. There is one alternative as regards pinions, namely the shaping of teeth either on a bench shaping-machine or by adapting a robust lathe to the method; but this lies outside the scope of this book.

Fly-cutting is a very good method of making

clock wheels. The cutter can be made to conform closely to the international standards on gear-cutting and it cuts rapidly if rotated fast enough, with few of the vibration problems that occur in gear-cutting on the lathe or miller. The fly-cutter itself is made to the cross-section of the space between adjacent gear teeth, rotates at high speed and traverses through the blank to leave a correct form on either side of the tooth space. Clearly, two such spaces should create a properly formed tooth.

Figs 161 and 162 show how this tool can be made using the lathe. Note that the tool-bit is of silver steel, which means that it can be machined in the soft condition and hardened with very simple heat treatment. The heat generated in cutting is largely passed to the chip, the small amount in the tool is dissipated in the large arc of rotation when the tool is not working, so that temperatures do not reach the point at which carbon steel starts to lose its hardness. Incidentally it is not generally realised that carbon steel is slightly harder than high-speed steel. The reason for the latter's wide use in machining is that it retains its hardness well above the temperature at which carbon steel becomes unusable as a cutting material.

The tool-bit is carried in an off-centre slot or hole so that, after turning, it may be reversed and gains machining relief from the simple geometry of the holder (Figs 163, 164). I have used cuttings speeds of up to 488m (1,600ft) per minute in 70/30 and free-cutting brass, this speed being merely the limit of my machine and the size of cutter that I could handle. I believe that speeds of at least twice this figure would be successful and give an even finer finish. The finish from fly-cutting is better than that obtained by milling the form with a multi-tooth cutter in high-speed steel, but only if the high speeds that are possible can be used. Unfortunately the technique is not successful on pinions.

Wear safety goggles when fly-cutting; it is even more important than in more normal machining.

Hand-cut wheels

Clocks whose gears were cut by hand should have replacement wheels made by the same techniques, in addition to which, if you do not have machinery that can be adapted to gear cutting, the only solution is to cut by hand. It is not a dreadfully skilled business, nor is it a long-winded one when you have polished up your skills.

Make a gear blank with root diameter, pitch diameter and outside diameter scribed on its surface; clear centre-lines across it and a hole cut in the centre for mounting on a collet afterwards (Fig 165). The centre-lines will tell you if any error came about as the bore was drilled and they will also enable the crossings-out to be made neatly. Free-cutting brass, or engraving brass, is best for blanks; obtain it in the hard or half-hard condition.

Take a large sheet of paper, mark a centre and then draw out a large circle four or five times the diameter of the blank. Using compasses or dividers, divide the large circle into as many equal parts as the teeth on the gear. This will be a matter of trial and error, but the error can be reduced considerably by tackling the job in the right way.

First of all draw a line from one side to the other (a diameter). If the count of teeth is to be an even one, the divisions that you determine can be set around the circle from each end of the diameter. Deal with one semicircle at a time and, when the dividing has reached half-way around this, move to the other end of the diameter and begin again there so that the divisions meet at the half-way point; any error when the two sets of divisions meet can be averaged over the last two or three marks. Repeat the method on the other semicircle (Fig 166).

If the count is an odd one, discover the setting for your dividers by trial and error and then straddle the centre-line at one end with a division and proceed as before, using these two points where you used one end of the diameter before. The other end of the diameter forms the other starting point, of course.

For counts divisible by four, it is a help to using drawing instruments to give an accurate quartering first, and for counts divisible by six one can make use of the well-known method of constructing a hexagon – the radius that drew the circle divides the circle into exactly six parts.

DIVIDING THE BLANK

Pin the blank over the paper, on a board, and line up its centre-lines on the centre-lines of

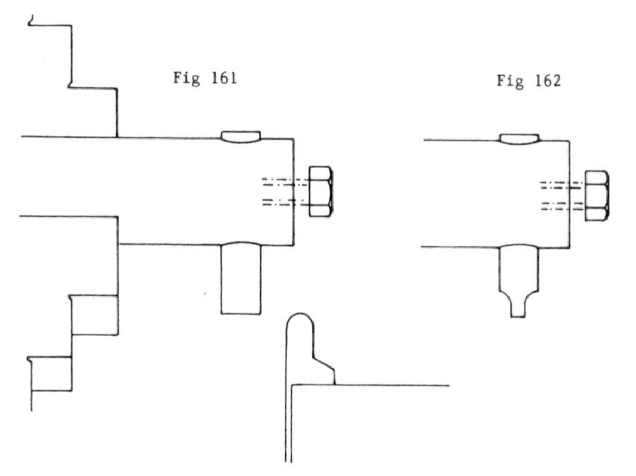

Fig 161

Fig 162

Making the fly-cutter on the lathe

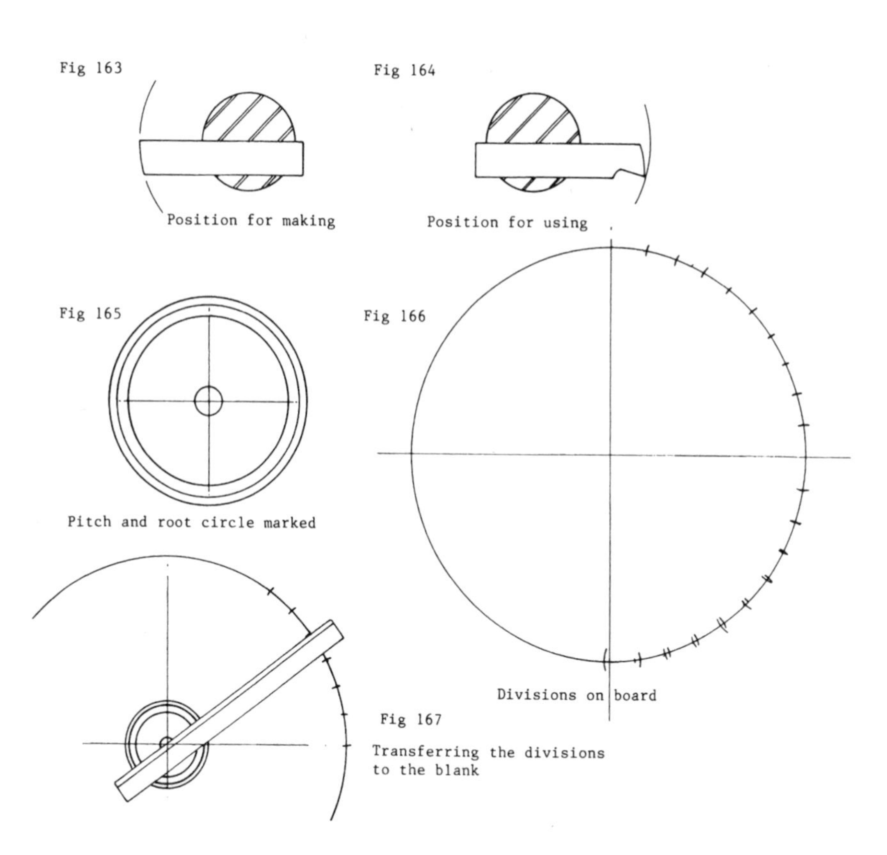

Fig 163

Position for making

Fig 164

Position for using

Fig 165

Pitch and root circle marked

Fig 166

Divisions on board

Fig 167

Transferring the divisions
to the blank

Fig 168a

Fig 168b

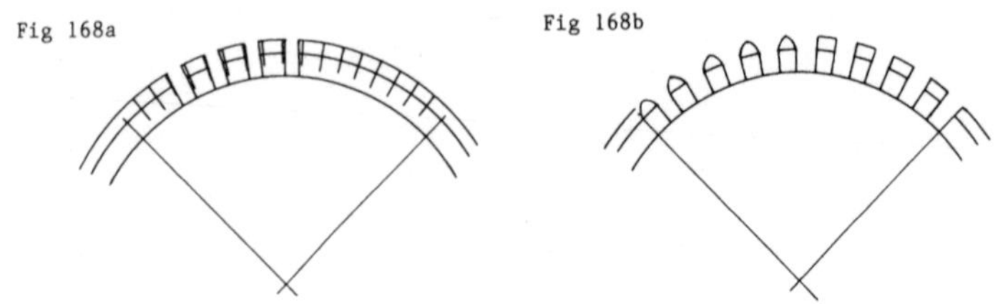

Cutting a wheel by hand

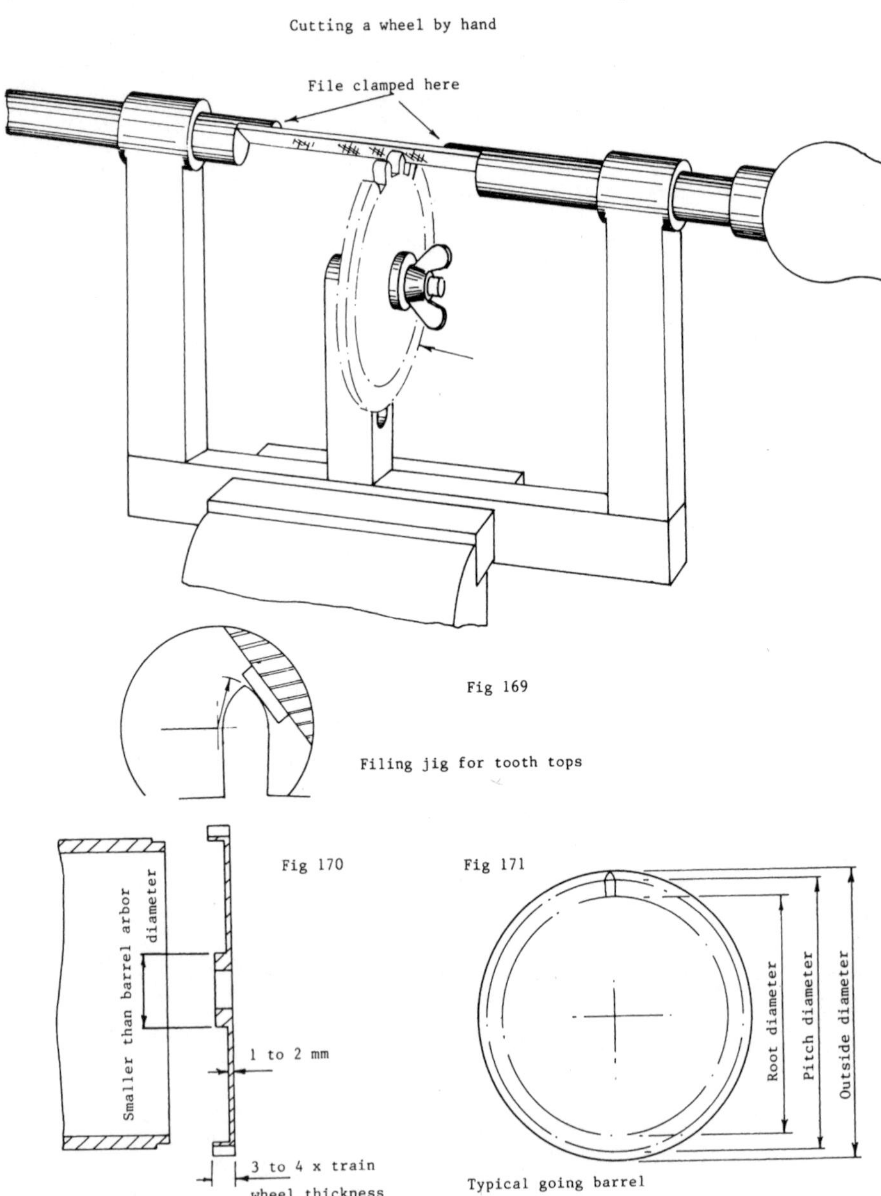

Fig 169

Filing jig for tooth tops

Fig 170

Fig 171

Typical going barrel

the large drawing. Now a straight edge and scriber can be used to line up the outer divisions with the centre, and scribe marks onto the brass blank (Fig 167). The whole point of this method is to ensure that all marking and division errors are made smaller by working inwards from a large circle to a small one.

When the blank is marked, place it between smooth jaws in the vice and use a saw to cut between the marks, down to the root diameter. The cut must be no wider than half the pitch of the divisions, as measured at the pitch diameter. Cut every tooth space in this manner (Fig 168a).

Now use a flat file with one side ground smooth, to widen these slots and make them parallel sided. Balance the amount of work that is carried out on each side of the slot, to keep the tooth centred on the scribed line of the divisions. When the space is of the required width, ie half the pitch at the pitch diameter, the tops of the teeth can be rounded (Fig 168b) using the same techniques as we used on the hand-cut pinion (page 91). A little more care must be taken, because thin brass is filed more rapidly than the steel of a pinion head and the tops of the wheels' teeth do not have as much space between them. (The angle between tooth centre-lines is less.) Fig 169 shows an old tooth-filing tool.

For guidance in the selection of gear-cutting methods, rough dates for the three methods of wheel cutting are: up to 1700 entirely hand-cut, 1700 to 1770 the tooth space was cut with a tool rather like a fine circular saw and the tops of the teeth were rounded using a file either jig-guided or, quite often, hand held; after 1770 most wheel-teeth were cut by formed cutters.

Spring barrels

This is a very frequent repair. There is rarely enough space under the root of the tooth for inserting teeth, and the only really satisfactory repair is to make, or have made, a new barrel ring. Having one made is not expensive, it costs just over twice as much as having an ordinary train wheel cut, the arbor is usually polished at the same time to match the new pivot hole and the barrel is then as good as new. Drilling hard, or even soft, steel pegs into the gap left from the breakage is a method which puts greater wear onto the pinion and passes the buck to the next repairer, probably with the additional need of replacing the pinion.

To do the job oneself, the old ring is sweated off (the solder is melted) or turned off, after the following dimensions have been taken: tooth count, outside diameter, tooth width, thickness of the end or the disc of the ring, diameter of the boss, diameter of the unworn pivot hole (Figs 170, 171). Again, actual gear-cutting techniques are outside the scope of this book; after soldering the new ring in position it is imperative that the flux be washed away and neutralised by a bicarbonate of soda solution. Otherwise there is a strong risk of corroding the spring when it is re-installed.

GEAR FORMULAE

These formulae are applicable to modern gear forms, and were not in being at the time that most of the clocks that we are concerned with were made. Nevertheless they are very useful in arriving at the dimensions of replacement gears. Metric measurements alone are used because the most commonly available cutters are based on the metric-module system. First of all we should make a few definitions.

P Pitch diameter
D Outside diameter
M Module
C Count

The relationship between the count of a gear, its module and diameter are defined by two formulae:

$M = \frac{P}{C}$ and $D = M(C+f)$ where f is a correction factor

Correction factors vary according to the number of teeth in a gear; the values of f for wheels and pinions are as follows:

	leaves	*f*
Pinions	7	1.71
	8	1.71
	9	1.71
	10	1.61
	11	1.61
	12	1.61
	14	1.61
	teeth	
Wheels	24	2.76

Wheels of less than 24 teeth do not often appear in clocks, neither are pinions of greater than 14 common. If you have such a gear to cut it would be as well to consult a specialist booklet on gear cutting. The factor is not solely defined by the count – in almost all clock trains the wheel is driving and the pinion is driven. You cannot, therefore, use the above table and extrapolate for a gear between 14 and 24 teeth.

Applying the formulae

Since, as already stated, we are usually concerned in repairing clocks that were never built with the above formulae in mind, a little care must be used in applying them. The important diameter is the pitch diameter, since this is where contact takes place. When it is merely a case of replacing a wheel or pinion that is worn or damaged, there will be very little difference between measuring the outside diameter of the original and cutting the desired number of teeth into it and calculating the pitch diameter first and then the proper outside diameter from that. Frankly, the latter is hardly worth the trouble unless the gear is very large and the train made with the precision of a very good-quality regulator.

A wheel or pinion that is missing presents a different problem. Measure the outside diameter of the remaining gear and, using the formulae, calculate its pitch-circle diameter (P_1). This diameter must be halved and then subtracted from the distance between centres (A), the result, when doubled, should be the pitch circle (P_2) of the missing gear.

$$P_2 = 2\left(A - \frac{P_1}{2}\right)$$

If two gears are to mesh properly they should have teeth that are, at least nearly, the same size as each other. It is just as well then, to make a check on the size of the teeth of the two gears in this calculation by dividing the pitch circle of each by the relevant number of teeth. The result is nearly always as you would expect, correspondence within 0.05 module, but occasionally the result is so far out that it is clear that either the centre distances have been altered or, much more likely, the existing gear is a foreigner introduced by a previous owner to make the amount of damage look less.

Incidentally, when carrying out the calculation for discovering the pitch circle of the existing gear, you will be presented with a figure for the module that is anything but nominal. Do not worry; take the value to two decimal places, calculate the pitch from this strange module and the outside diameter of the new gear, and then use a nominal size for cutting. Figs 172, 173 and 174 show gears that are, respectively: too close, correctly meshed, and too far apart.

Lantern pinions

The teeth of lantern pinions are in the form of rods. At one time these were very popular in cheaper movements and, as a consequence,

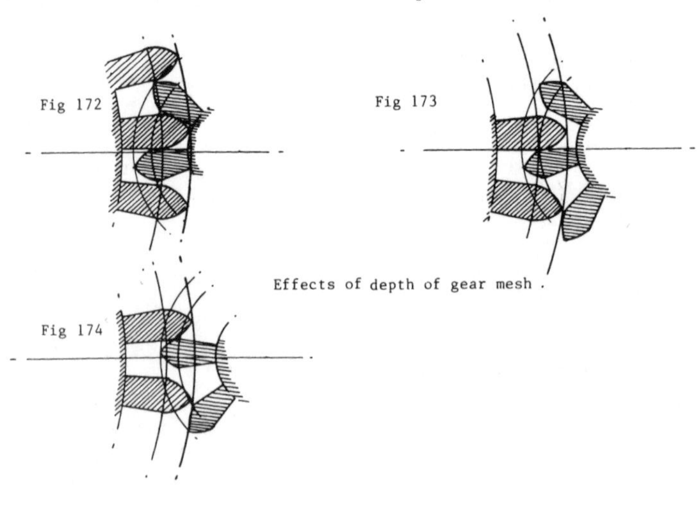

Effects of depth of gear mesh.

Fig 175

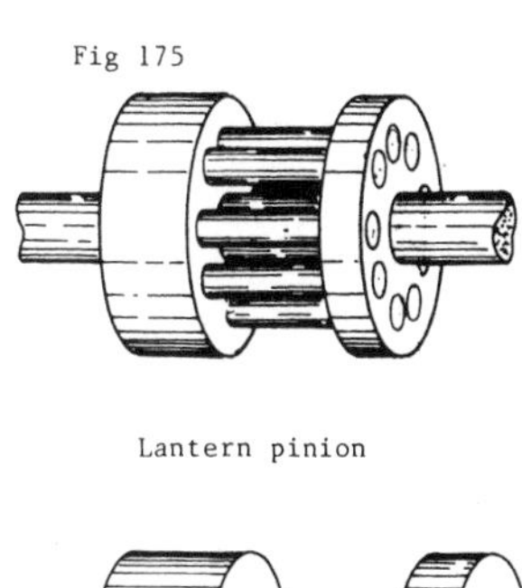

Lantern pinion

Fig 176

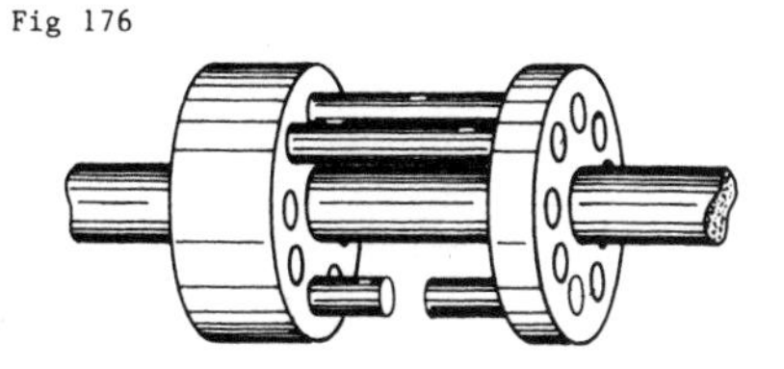

Trundle cut for removal

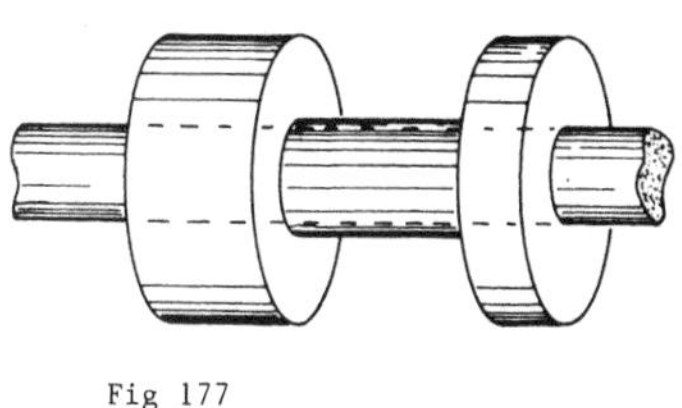

Fig 177

Spool for new pinion

Fig 178

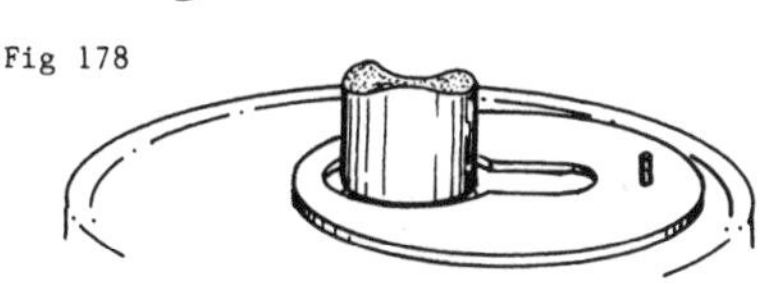

Pinning of keyhole washer

clock repairers have tended to look down on them. In fact the type is extremely good, and it is a pity that mass-produced movements cannot make use of them today, but apparently it is cheaper to use cut pinions in a rough, slightly tough, condition (Fig 175).

Repair of a lantern pinion is best carried out with pivot steel – polished hard rods that are available in a wide range of diameters tempered to blue temper. The original rods (properly called trundles) are held in by swaging the thinner brass collar so that the holes are pressed over the trundle ends. Usually this is fairly lightly done and the rods can be removed by filing the face of this collar and then sliding the worn, bent or broken piece along and out. If this cannot be done, use cutters to cut a section out of the middle of the trundle and then extract the two remaining pieces through the space between the two collars (Fig 176). A drill can then be used to open the clenched-over holes so that new trundles can be inserted. The latter must, of course, be the same diameter as the old ones, and they must also be short enough for the end to sink beneath the surface of the thinner collar as the old one's did. Lock them in place by either holding the collars in the chuck and swaging the holes over, or by using a modern adhesive such as Loctite.

Total replacement of a lantern pinion is sometimes necessary if the collars have been damaged in an accident or by an inadequate repair. Rather than make separate collars again, I prefer to turn a bobbin-shaped piece of brass that supports both ends of the trundle (Fig 177). Make sure that the small diameter in the middle of the bobbin allows space for the tops of the gear teeth to enter and clear out the dirt and old oil that may accumulate after it leaves your workshop.

Tower-clock pinions are sometimes made as lantern pinions. These are often large enough to allow the trundle to rotate in its bearing, which has the advantage of presenting new wearing surfaces to the wheel teeth as the movement operates. It also tends to shed a large part of the abrasive substances that tower clocks are prone to. Nineteenth-century clockmakers disparaged the use of rotating trundles, pointing out that they merely displaced the point of friction, without decreasing it. The obvious point that friction may then be applied on a shorter moment arm and, more importantly, that friction may be confined to an area where lubrication is more readily controlled, seems to have been missed. A modification in detail of this type is perfectly allowable on a clock that has to work in such poor conditions as a tower clock.

Great wheels

Longcase clocks and some others have great wheels that are kept in place by means of a 'keyhole' washer (Fig 178). The arbor has a groove machined into its diameter and the washer has a diameter equal to the outside diameter of the arbor, cut off-centre, and another equal to the root diameter of the groove on the centre-line, thus creating the keyhole shape. The washer is slid over the end of the arbor and then pressed sideways into the groove so that the wheel is kept in place. To stop the washer coming free accidentally it is either pinned or fastened with a screw. The latter gives no difficulty when one is removing the wheel, but the pin is frequently very hard to pull out. It should be left protruding by about 1.52mm (0.06in) so that it can be pulled free, nevertheless this is not usually a successful exercise. It is better to either convert to a screw, or put a pin in a position where it can be driven right through the hole and into a clearing-hole in the end of the barrel. Since the barrel probably has a hole in it at this end to accept the knotting of the cord, this arrangement is therefore quite simple.

Moving wheels

It is sometimes possible to move a wheel along its arbor so that it will run on another part of the pinion. This is an acceptable repair, but it must not be done by dishing the wheel. The whole collet can be moved, but this usually results in an eccentric wheel. Either remove the wheel and face the shoulder of the collet back – always supposing that this is the direction that you wish to move it in and that the shoulder of the collet is thick enough – or make a new collet to the old form and re-mount the wheel. If the wheel itself is worn it should be reversed to use the unworn side of the teeth, and any pins on its face moved to cope with the new arrangement. This cannot be done with hoop wheels, of course.

9
Suspensions and Crutches

Christian Huygens is credited with having invented the crutch, pendulum and pendulum suspension in one design of clock in 1657, though there are other claimants. Prior to the pendulum, the movement of a clock was governed by means of a foliot or a balance wheel, the foliot being a bar pivoted half-way along its length with weights arranged variably along the resultant arms. If the weights were moved out from the centre the device tended to oscillate more slowly, if they moved inwards it oscillated faster. Both balance wheel and foliot were supported on a vertical spindle with conventional pivots, the lower one having an adjustable end-plate. The spindle carried the verge flags and was usually termed a 'staff'.

The changes that were brought about by installing a pendulum resulted in the verge staff operating in the horizontal plane, and some provision being made to support the pendulum. Huygens' is the earliest system illustrated – the staff is fitted with cylindrical pivots at each end, and it carries a crutch that transmits the impulse to a pendulum that hangs from cords. This is virtually the same system as that used in pendulum clocks today. There is little weight operating on the pivots, just that of the verge staff and the crutch.

KNIFE-EDGE SUSPENSIONS

Strangely enough, the clocks made in this country that employed the short pendulum and verge escapement ignored the crutch and attached a small pear-shaped pendulum bob to the verge staff itself. The weight of this bob seems to have persuaded the makers that a cylindrical pivot would offer too much friction and so a knife-edge bearing was installed at the pendulum, or back-cock, end (Fig 124). The knife-edge is often made knife sharp, which leads to difficulties as it has a tendency to chip if left dead hard and to turn over if tempered. There is absolutely no reason why this edge should not have a slight radius; indeed I believe the path of the bob is thus brought a little (very little) closer to a cycloid, with consequent slight lessening of circular error. The main point, though, is that a blunted knife-edge is no worse as regards time-keeping than a sharp one, and better mechanically.

The best arrangement for a knife-edge suspension is that of a wedge against a flat hard surface; unfortunately this makes it possible for the verge staff to move to left or right of true centre as a result of a slight shock, or continuous vibration. Restraint of some description has to be used. Banking pins (an obstruction at either limit of required motion) are, as has already been mentioned, not a good idea; they add a variable degree of friction, or provide a fulcrum for the knife to kick against. The usual solution is to drop the knife-edge into a wide-angled notch with a narrow flat at the apex, or set the knife into a concave bed (Fig 140); in both cases gravity seats the wedge in the centre and keeps it there. However, this tendency to wander plus the restraint constitute a large part of the escapement error (to use the term a little inaccurately). Both elements of the suspension should be hard, but I can see little virtue in tempering either, so long as the knife has a rounded edge.

Knife-edge variations

There are one or two variations on the knife-edge, the most common being the Black Forest suspension which consists of two wire U-pieces, the upper one fastened to the frame so that the base of the U is horizontal, and the other having two looped ends that pass over the base of the first (Fig 179). The pendulum rod hooks onto the lower one. A very frequent cause of failure to 'go' in old Black Forest clocks, lies in the wear that takes place between these two pieces of wire. As soon as one wire begins to bite into the other, the contact is no longer point contact but along close-fitting curved surfaces generated by the wear. Spreading the arms of the lower 'U' will often cure the problem by moving the looped ends out of the worn grooves, but if this cannot be done, because of the room available or the condition of the wires, replacements for either or both can be made from pivot wire of similar diameter. Anneal the pivot wire first, polish the surface after this and then bend into the form of the original parts.

Another variation appears in the 'Keeless' gravity clock of the 1920s (Fig 180). In this case the clock has a compound pendulum (two bobs with the pivot between them), and beats 60 to 80 times a minute. The pivots for this pendulum have a plain cylindrical bearing of small diameter, that lies in a bearing hole some five or six times larger. Any failure of the suspension is as a result of deterioration of the pivot or pivot-hole surfaces. The small pivot can be burnished in the lathe in the normal fashion (see Chapter 4), and the larger one can be burnished with a hard polished-steel rod. However, any measurable removal of metal from either will alter operation of the clock. Take careful note of both diameters before starting, so that in the event of having to do more work than expected, the geometry can be restored.

Verge knife-edges do not necessarily have to lie on the centre-line of the arbor (they can be made deeper and so, stronger if they do not); but there are two things to bear in mind if the edge lies below the centre-line. Firstly, the mass of the staff or arbor will become part of the pendulum, making the period of oscillation longer than indicated by the apparent length of the rod. Secondly, that the action of the verge flags must be determined with reference to the knife-edge, rather than the centre of the arbor (Fig 181).

SUSPENSION COCKS

Apart from those clocks already mentioned, most pendulums are hung from a suspension cock of some kind. Few failures can be attributed to the back or suspension cock, in a clock that has been running for some years, and if it shows no sign of having been weakened by age or modification. Inspect the mounting of the cock, if it is in the English style (Figs 182a and b) and fastened directly to the back plate, there are two possible failings. The most obvious is the fitting of the suspension top block; it should be free, but not rattling. Less easy to spot are small pieces of grit, raised metal or anything else that prevents the mating face of the cock being held flat against the back plate of the movement. If the cock is fastened to the back plate, and then both are held up to the light, no gaps should be showing between the two parts.

Many American clocks use a simple suspension cock made from a short cylinder of brass riveted into the front plate (Fig 183). Check that there is no movement either up and down, or side to side; any movement can be remedied by using a light ball pein hammer on the rivet. If, for some reason, the movement is not to be dismantled, a drop of a cyano-acrylic adhesive will achieve a good repair.

The Brocot suspension has an inbuilt adjustment that varies the length of the pendulum at will, to alter the timekeeping of the clock (Fig 185). Most are held by one screw only, and prone to damage on the mating surface as a result of careless assembly. Test in the same manner as for English back-cocks. Other faults in this type are really in the domain of the suspension itself and will be dealt with later in this chapter.

Regulators should have their suspension cocks mounted away from the movement and on the back of a rigidly supported case, or directly onto the wall behind. Other than that their faults are the same as those of the normal longcase or wall clock with a back-cock; check that the mating surfaces of cock and mounting show no spaces.

Silk-thread suspension

Huygens' intended suspension was a cord and this method can be found in French and other small mantel clocks. It also appears on early Dutch clocks, particularly Salomon Coster's, Huygens' clockmaker (Fig 186). The term usually applied is silk-thread suspension, but the suspension cord may just as usefully be a mercerised cotton thread.

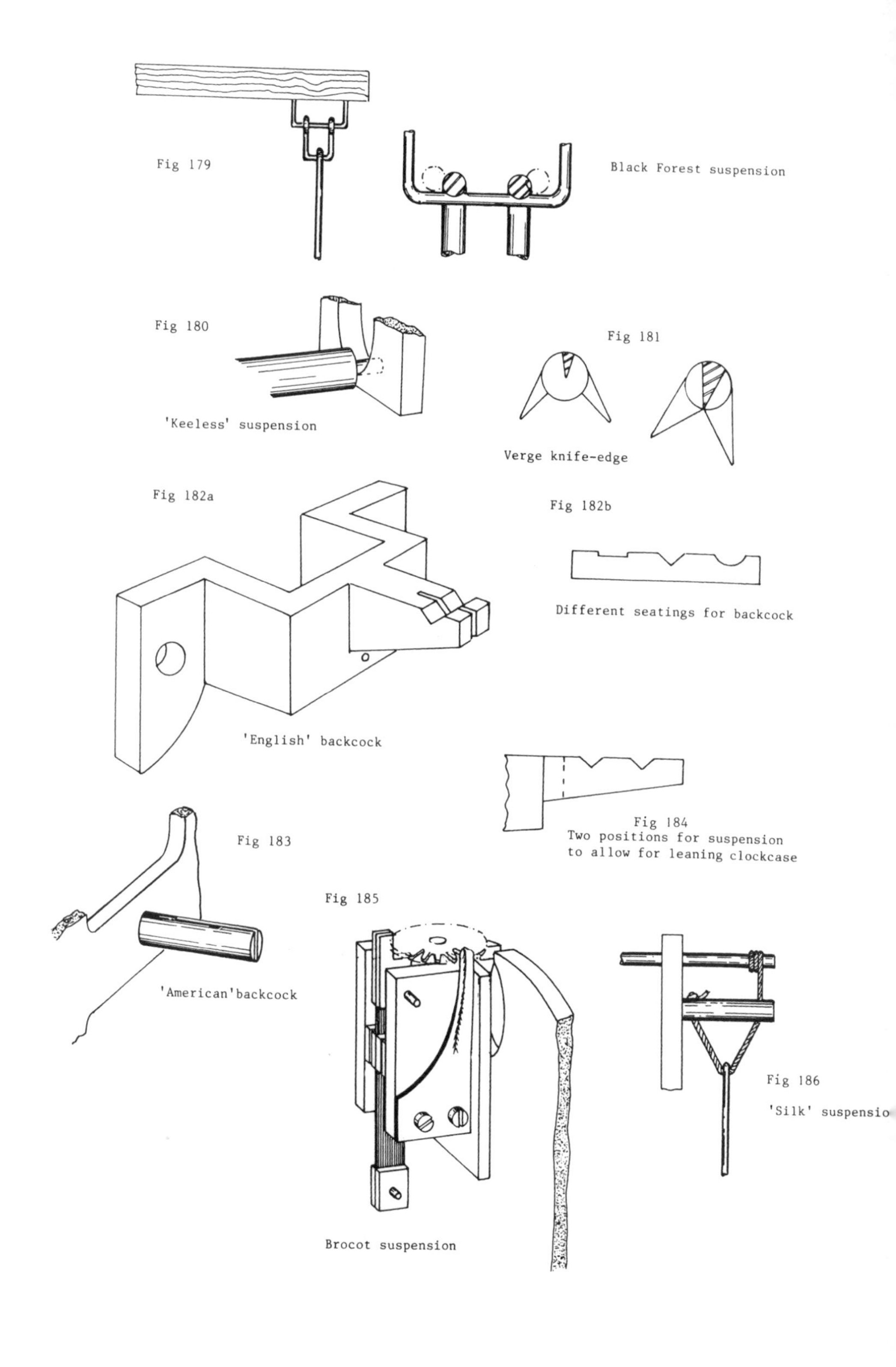
Fig 179
Black Forest suspension
Fig 180
'Keeless' suspension
Fig 181
Verge knife-edge
Fig 182a
'English' backcock
Fig 182b
Different seatings for backcock
Fig 183
'American'backcock
Fig 184
Two positions for suspension
to allow for leaning clockcase
Fig 185
Brocot suspension
Fig 186
'Silk' suspensio

The thread is held in a loop from a suspension cock, and the light pendulum hooks over the loop. One end of the thread is wound, windlass fashion, onto an adjusting arbor that is frequently accessible through the dial. Inspect all surfaces that the thread can rub on. The holes in the suspension cock should, of course, be smooth after many years of use; but the hook of the pendulum is often marked by the pliers of previous repairers. This is one of the few clocks that should not have the pendulum removed for transport; if a clip is provided, make use of it, otherwise fit a light rubber-band and keep tension on the suspension.

There are few things that can go wrong with a silk suspension. The arbor that winds the thread up or down for adjustment is a friction fit, and may work loose. It can be tightened again by either tapping the end of the arbor-bearing with a polished and slightly convex punch or, if the mounting is removable, pinching in the outer diameter. Please do not score the arbor with either pliers or chisel. The thread may be 'fluffy'. Throw it away or wax it smooth again. If the thread is in any way stiff, throw it away anyhow. It is not my normal manner to be so cavalier with parts of old clocks, but a bad thread can cause a lot of trouble on this suspension and I can see no reason to keep it.

Engagement with the crutch is as important as in any other pendulum clock, and we will deal with this later.

Spring suspension

The most common form of suspension is by a thin spring. There are usually three components; top block (for attachment to the cock), bottom block (attachment to the pendulum rod); and the spring itself. Fig 187 shows a variety of these.

Failure of a spring can be by way of:

1 The fitting of the top block to the cock. It must be firm; at the same time, in anything but a Brocot-type suspension, it should be capable of being swivelled by the weight of the pendulum so that there is no chance of it having to work at an angle to the path of the pendulum. If a suspension is stiff in the suspension cock, it is quite possible for it to stick out at an angle to the vertical and impart a sideways motion to the pendulum, causing it to 'roll'. This is a common fault in mantel clocks with light pendulums, particularly the Brocot mentioned previously. It is a worthwhile safety measure, when setting up a clock, to give a light tug downwards to the pendulum to align it to true vertical.
2 The fitting of the bottom block to the rod. This must be firm, with no rattle.
3 The spring must be flat, with no cracks, creases or bend. Quite obviously some clocks can get away with conditions of spring that others could not use at all. The longcase with its heavy pendulum will often go even though the spring has a permanent set that curves it, or twists it, or has a crease running across it. Unless there is some good reason for preserving a particular spring (it may be the original hand-made one), change it at the first hint of any mark on the surface whether it be a crease, crack or corrosion. It is not sufficient simply to use emery cloth to rub the corrosion marks away. Hand-made springs on antique clocks can often be preserved, if one takes into account the small amount of spring that is actually working. In most longcases the spring may be anything up to 100mm (4in) long and yet the working piece can extend for no more than 12.5mm (0.5in) at the most below the suspension cock. Most damage to a spring occurs at top or bottom and the operation of the clock will not suffer if it is shortened. The position of the bottom block will change, of course, and if it fails to coincide with the crutch forks some other adjustment must be made. It is at this point that you must judge whether the crutch can be shortened without damage to the appearance of the clock. It is in fact a choice between preserving the spring, or shortening the crutch, and you must make your own mind up as to which will serve the clock best.

Replacement of spring suspensions can be made by checking the original against the illustrations in a supplier's catalogue. Most companies provide full-size illustrations for you to match.

When attempting a match the most important things to duplicate in the new spring are the mounting to the cock and the cross-sectional dimensions of the spring. Very

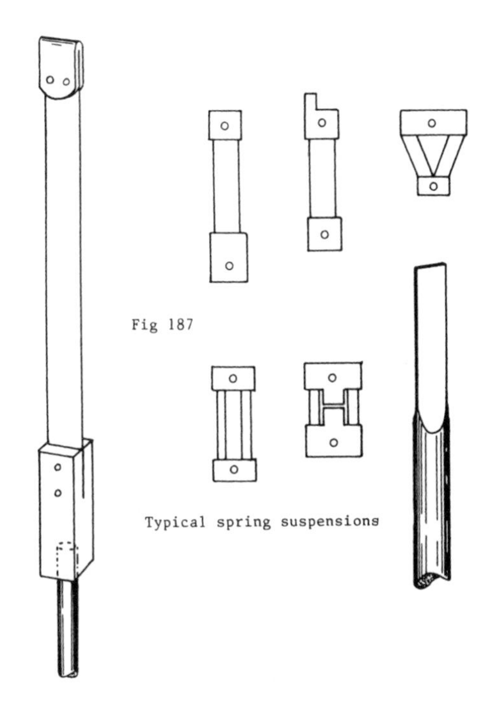
Fig 187

Typical spring suspensions

frequently small suspensions have two arms separated by a gap; this is to give them stability, and control the tendency of the pendulum to roll or change its path as the earth rotates. (A pendulum hanging by a single thread will apparently change the path of its swing, because the earth is rotating in space and the pendulum is not.) If the only replacement available has a wider space between the arms, and all other requirements are met such as the shape and size of the blocks, width of the individual arms and approximately the same length, accept the replacement. A little extra length will not matter to anything but the location of the crutch and the rating adjustment of the pendulum bob; and as long as there is about 5mm (0.2in) of spring length, most mantel and other small clocks will be quite happy. As was stated before, very little of the spring is actually working.

PENDULUM BOB SIZE

One consideration of spring suspensions has not yet been touched on – the strength of the spring. Will it be strong enough to support the weight of the pendulum for decades of working?

The standard spring supplied for longcase clocks is about 6mm (0.25in) wide and 0.18mm (0.007in) thick, and I have tested this for 10,000,000 reversals of a 10kg (22lb) pendulum bob without sign of failure. In the event, the clock that the pendulum was designed for had two such springs mounted with a space between for stability, but it does show that there is no possibility of the more normal bob weight of 1 to 2.5kg (2 to 5lb) breaking the spring during service. The amount of semi-arc (the pendulum swing either side of vertical), has a bearing on the stress developed in the spring. The semi-arc used in the above test was over 3 degrees which is around the norm for a longcase, or any anchor recoil escapement with the pallet pivot, and the flexure point – it is more of a line than a point – almost in line. Any clock which has the suspension hung higher than the pallet-arbor pivot will drive the pendulum through a smaller arc than this. Problems concerned with the stiffness of a spring are more often those of too much stiffness rather than too little.

Bulle suspensions

The Bulle clock is electrically operated by a 1.5 volt battery, and the suspension is made of non-conductive material. Its top and bottom blocks are of brass, and the flexible 'spring' between is a tape (Fig 188). Since these are not generally available, you will almost certainly have to make one yourself if a replacement is needed. The most available material for this is ribbon, of the same width as the original, and with a selvedge along both sides. These selvedges are often raised above the level of the ribbon, thus the brass blocks mainly grip on the ribbon's edges. Make sure that the ribbon is not in tension when the blocks are screwed together at top and bottom, otherwise there will be a tendency for the sides to draw together.

Pendulum rods

The stability of a pendulum can be affected by the rod. In the case of a seconds pendulum, the rod is made stable by fastening it firmly at the end of the spring suspension so that it can only move in one plane, its swing. However, some pendulums consist of a rod that is only hung from the bottom of the suspension (Fig 189). This makes it easier to

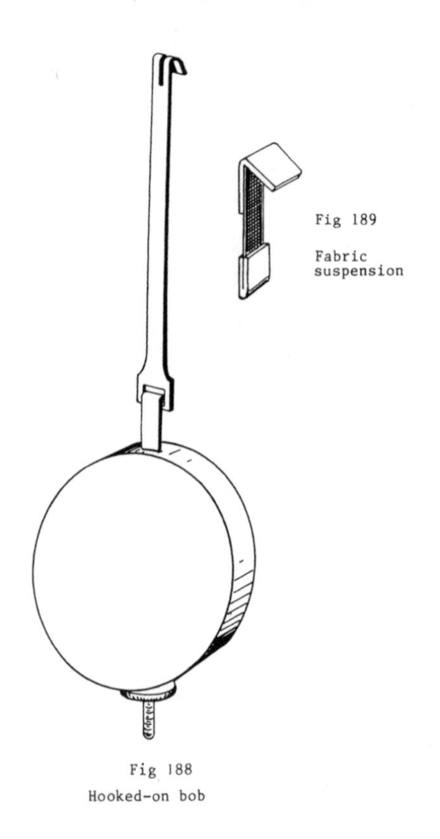

Fig 189

Fabric suspension

Fig 188

Hooked-on bob

remove and rehang the pendulum bob, and it also prevents the pendulum damaging the escapement when disturbed by winding the clock, or even moving it. Although the pendulum bob is not therefore fastened directly to the suspension it must still be allowed movement in one plane only, and this is usually managed by making the hook, and the part it hooks onto, out of flat strip so that the width of the strip controls any tendency to swing outside the plane of the pendulum beat. Ensure that the parts that engage, do so across the whole of their width. If, for instance, the one part is a horizontal support over which a flat hook fits, engagement at one side and not the other will give a very unstable contact. Many German clocks seem to contradict this requirement in that the bob is free to swing at right angles to the arc of the pendulum, but in fact the strip of metal that forms the hook makes contact over its entire width, and prevents any twisting or rolling of the bob.

Suspension top blocks

Several methods of holding the suspension in the suspension cock are shown (Figs 190, 191); however it is done the suspension must be allowed to align itself with the pendulum. In most cases this is simply the matter, already discussed, of allowing a certain amount of freedom so that the weight of the pendulum pulls it into the vertical. One type of suspension does not allow this, namely the type of American suspension that is formed by rolling the end of the pendulum rod into a thin spring. If the spring is held in place by a taper pin, as I am sure was the original intention (Figs 191c and d), the pendulum attains plumb easily; but many of these suspensions that I have seen have been held by bending the top of the spring over the back-cock and this allows no movement at all (Fig 191e). This form of fastening should be corrected by inserting a taper pin, either above the cock or through it. These suspensions are still available from suppliers in this country and, presumably, the USA.

400-day suspension blocks

The suspension of the spring that supports the torsion pendulum is also the means of putting it in beat. The beat is adjusted by twisting the top block in the horizontal plane, forcing the pendulum to rotate more to the clockwise or counter-clockwise as needed. Cheap models use the same screw to hold the top block in the cock as to lock it after the setting in beat. To prevent the top block twisting when tightening this screw, make a small spanner out of brass strip and, once the beat is obtained, hold the top block while the screw is tightened. Better-made clocks have separate means of setting in beat and locking but, even so, it is simpler to use a small brass spanner or a brass blade (if a screwdriver slot is the means of adjustment), for the small movement that will be needed for final adjustment.

CRUTCHES

Engagement between pendulum and crutch must be close, but not in any way tight. As the pendulum swings there is a slight up and down motion of the rod in relation to the crutch, any tightness will therefore stop the clock. The movement is very slight, but positive.

The crutch can either fit around, or carry a

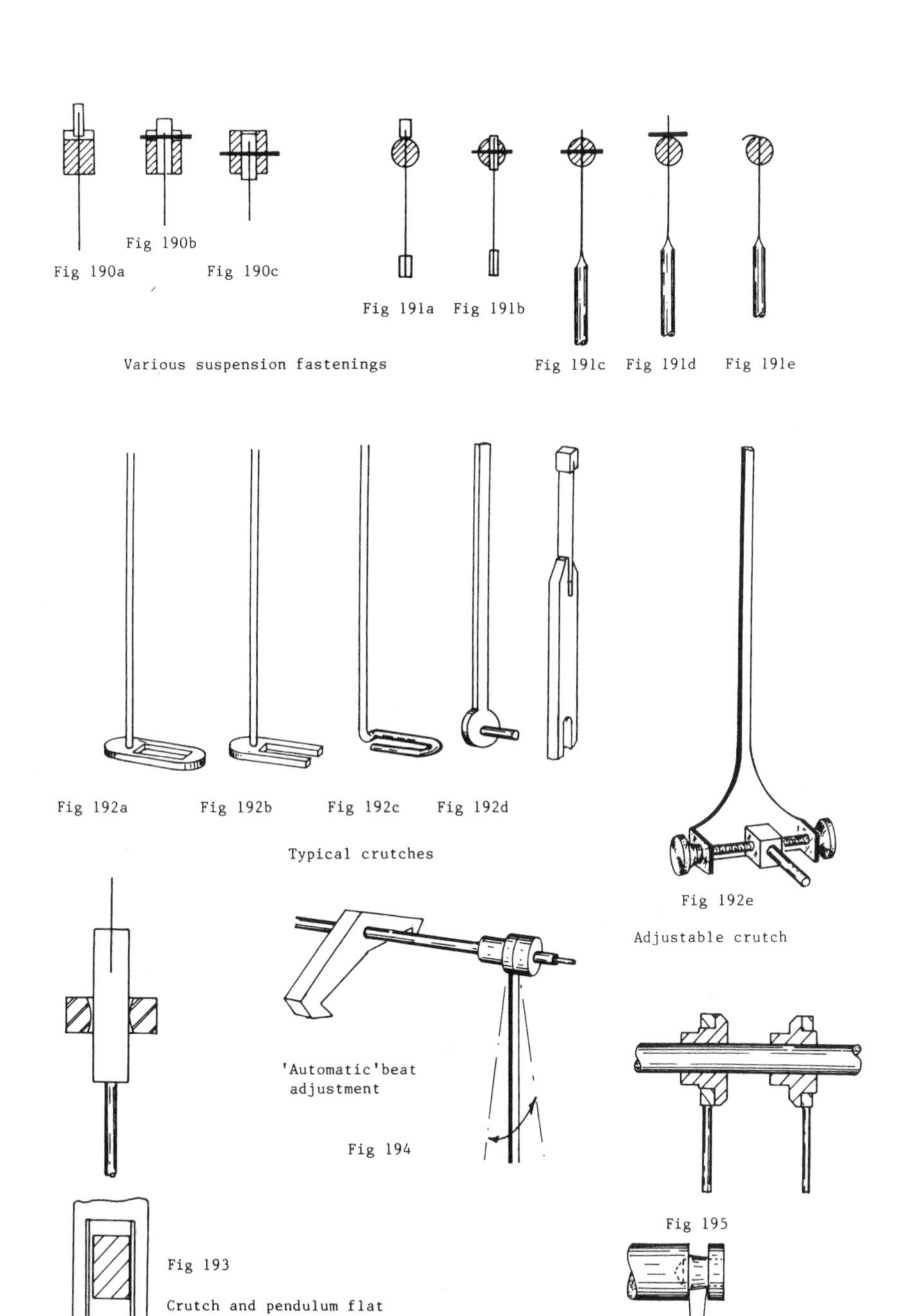

Various suspension fastenings

Typical crutches

Adjustable crutch

'Automatic'beat adjustment

Crutch and pendulum flat

peg that slides in a slot in, the rod or bottom block (Fig 192). It makes no difference which method is employed, one is as good as another, but it must be remembered that the crutch is often used to stabilise the pendulum in addition to the effect of the suspension. In the case of a fork there should be equal contact between fork and block at all points (Fig 193), otherwise the crutch will induce a rolling motion and not eliminate it. Equally, in the type with a peg and slot, contact must be through the thickness of the block or rod. If the plane that the crutch swings through is not parallel to the plane of the pendulum, there is a strong possibility of the pendulum rolling. Again, if the fork or peg is not truly at right angles to the crutch it will be putting a twist on the suspension, and rolling is very likely to result from this fault also.

Worn crutch forks can often be corrected by squeezing them together, and then using a file and crocus paper to put everything to rights. There is a limit to this, however, and if you appear to be the umpteenth person to apply this method do not squeeze in any further, change the bottom block of the suspension for a thicker piece of brass.

Peg and slot crutches tend to acquire a notch in the slot of the rod. File this out, make sure that the slot is parallel, smooth and square with the peg, then make a new peg, also polished, to fit.

PUTTING IN BEAT

The adjustments for beat are varied. The simplest is simply the bending of the crutch wire and this is very effective; but the wire must be soft or you run the risk of breaking the pallet pivots. If a longcase clock has been repaired with a hard crutch wire in the past, anneal it; it certainly was not hard when made and it is a liability now (Fig 192).

Many French clocks and modern clocks have a friction fit between the pallet arbor and the crutch (Fig 194). If this has been mangled so that it goes tight and then loose, remake it; it is far too much trouble to keep and try to repair. The critical part is the top of the crutch; this is either a brass washer or a short cylinder and its faces must be dead parallel to give an even friction fit (Fig 195). Make it by facing a piece of brass in the lathe and then parting off at the same setting. This is the only simple way to achieve dead parallelism. The washer then fits onto the body of the collet that attaches the crutch to the arbor and is swaged in place, or riveted with a light hammer and punch. Follow the original design.

Vienna regulators and true regulators have a more complex system (Fig 192e), but there is little that can go wrong; only hamfistedness can damage it and the remedy is quite evident. It will either be a matter of remaking the brass body or, more likely, making new screws.

The actual operation of putting any pendulum in beat is always difficult to describe, but I will try:

1 Move the pendulum to the right until a tic is heard.
2 Release the pendulum.
3 If there is another tic, go to 5.
4 If there is not another tic, move fork or peg to the right and return to 1.
5 Move the pendulum to the left, until a tic is heard.
6 Release the pendulum.
7 If there is another tic, the clock is in beat.
8 If there is not another beat, move fork or peg to the left and return to 1.

Many French marble clocks – the case is not usually real marble – are too heavy to make it convenient to keep turning them for beat adjustment. Another point is that the shelf or sideboard that a particular clock is sitting on may not be horizontal, and a clock will go out of beat again as it is turned back to its working position. There is a simple way to make the setting of these clocks more convenient. Leave the round movement a little slack when fitting it in the case, the pendulum can then be set roughly with its back towards you and, when facing you, can be put exactly in beat by twisting the movement by way of the bezel (the frame of the glass). When the clock is in beat and facing the proper way, turn it around once again and tighten the movement-holding screws, being careful not to move the setting and put it out of beat again.

10
Strike and Chime Work

Train wheels and pinions, pivots and pivot holes, behave in the same manner as those of the going side; they should be tested in the same way and will respond to treatment in similar fashion. The only difference is when a wheel or pinion suffers shock as the hammers strike or a sudden stop as the mechanism comes to the end of a strike, and this will be dealt with as the chapter proceeds. The simplest of all mechanical striking methods is the strike on passing.

STRIKE ON PASSING

This is a simple business of arranging for one hammer blow to take place every time the minute hand revolves. Almost all methods utilise a pin lift from the minute pipe, from the cannon pinion attached to the pipe, or from the compound gear commonly termed the minute wheel (Fig 196). Since the operation is merely that of gradually lifting the hammer tail over a period of time and then releasing it as the minute hand reaches 12, there is very little for the repairer to do. Examine the bearing parts and also the wearing parts of the lift for notching or corrosion, to make sure that the clock is not working too hard to lift the hammer. Apart from that, look at the positioning of the bell or gong and the strength of the hammer spring if there is one.

Because the lifting of the hammer is performed by the going train, the amount of necessary work must be minimised:

1 The spring should not put so much load on the going train that there is an audible change in the latter's working when lifting the hammer.
2 The hammer head should be so placed that it provides a large amount of the energy required for the strike by simply falling onto

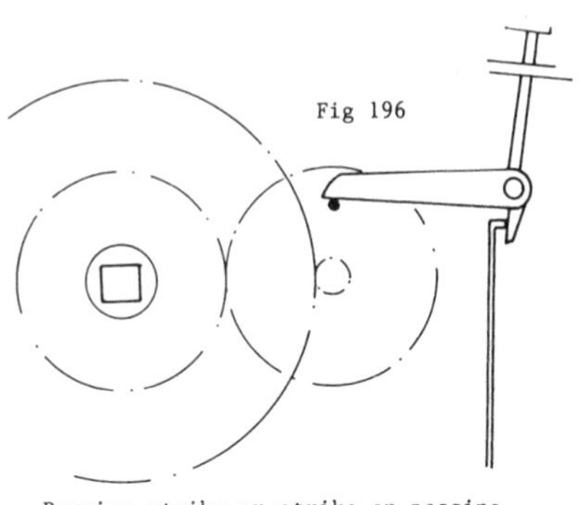
Fig 196

Passing strike or strike on passing

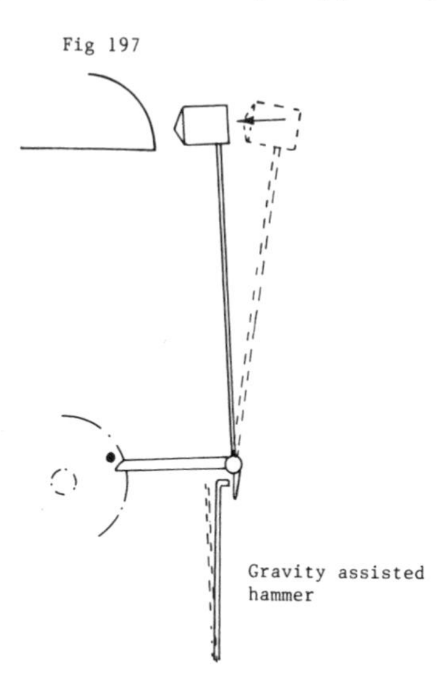
Fig 197

Gravity assisted hammer

the bell. The spring may then not be strained for anything but the last part of the lift and, when released, gives initial impetus only (Fig 197).

3 To achieve this the bell must be positioned so that the hammer can fall free before striking.

There is usually sufficient freedom in positioning bell and hammer to gain these benefits without modifying the movement in any way. If not, you must live with things as they are; it would be wrong to alter the original design in an elderly clock that has successfully worked for decades, if not centuries.

COUNT- OR LOCKING-PLATE STRIKING

This is the older of the two major striking systems and, despite the difficulty of synchronising hands and strike after any adjustment of the hands and the impossibility of arranging a repeater mechanism, it has remained in use to the present day.

Figs 198 to 215 show the most frequently seen layouts for the count-plate strike and how they operate. Essentially the system consists of a set pattern of striking progressed by the operation of the clock, the pattern being governed by a plate or a wheel with spaces that measures the number of strikes. There is no way of making any particular part of the plate repeat its strike and, if the hands are moved faster than the successive strikes can be made, the clock has no means of correcting itself so that the right number of strikes is made when next required. We should now take a look at the major sources of trouble.

Worn train and pivots

The same criteria apply as in the examination of the going train. All holes that are oval should be bushed, if this exceeds the amount that can be tolerated by the gears. If the gears still mesh well and give smooth transmission, the ovality has not affected the proper working of the clock.

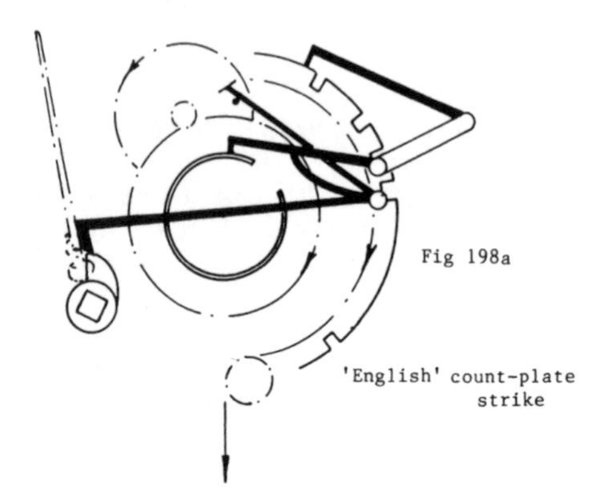

Fig 198a

'English' count-plate strike

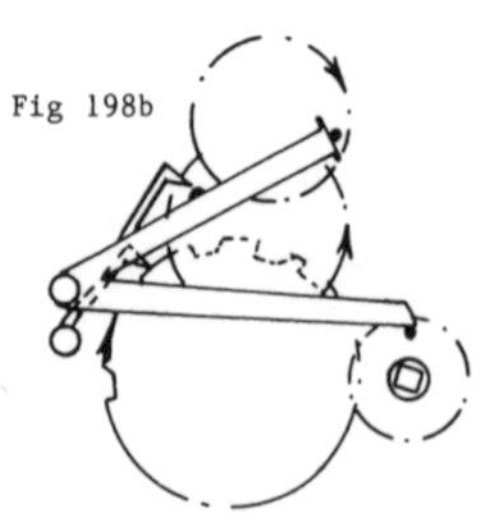

Fig 198b

French style of count plate

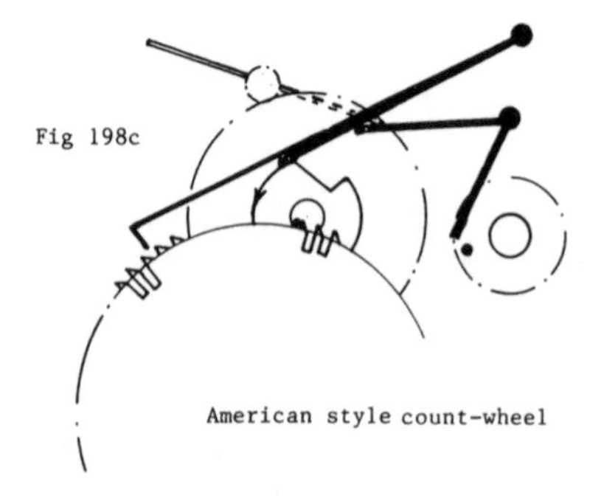

Fig 198c

American style count-wheel

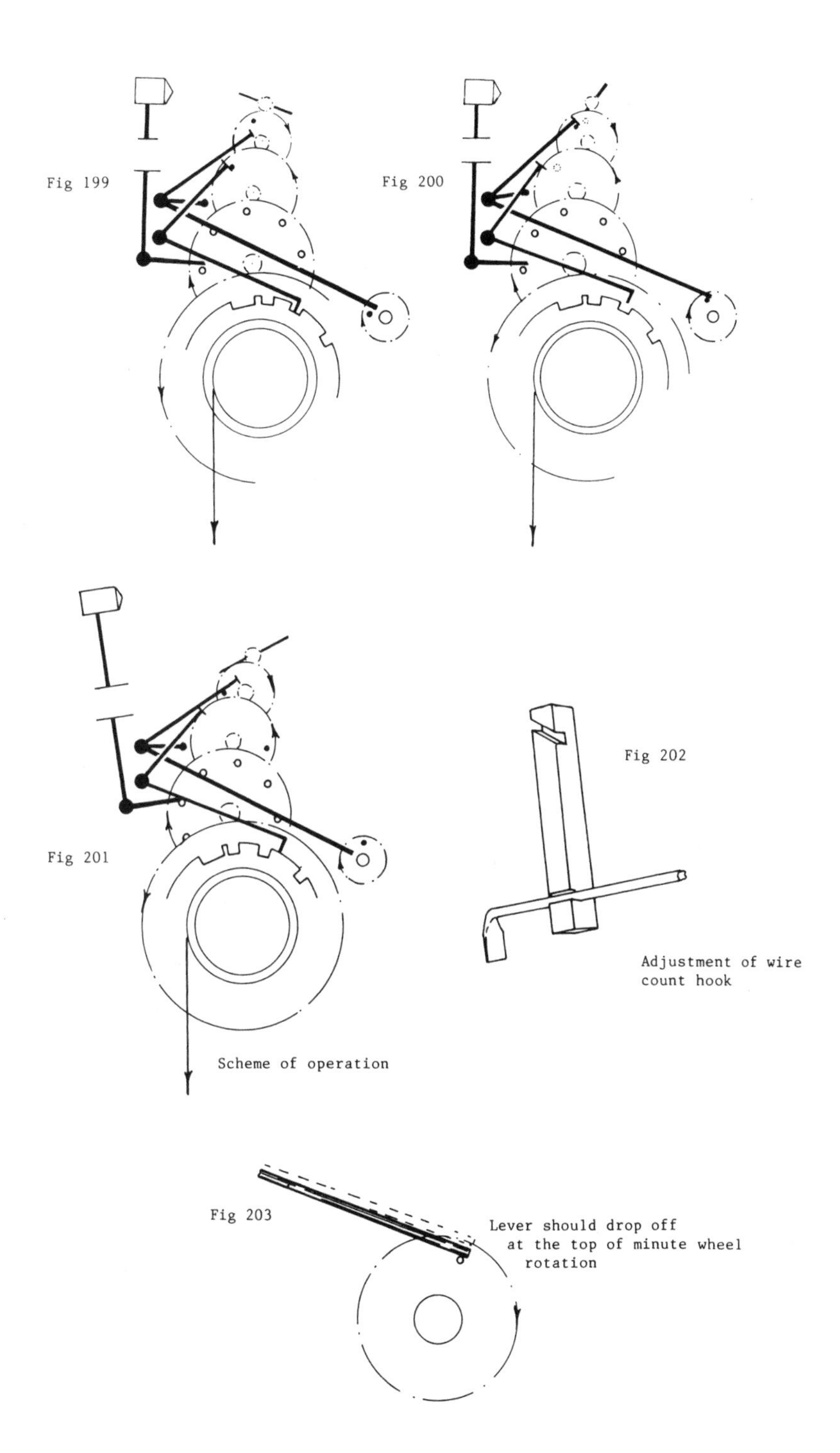
Fig 199
Fig 200
Fig 201
Scheme of operation
Fig 202
Adjustment of wire count hook
Fig 203
Lever should drop off at the top of minute wheel rotation

Lifting pieces

Lifting pieces in clocks with count plates tend to be fairly simple, mainly because the clock is normally a relatively cheap one and there is no desire on the part of the maker to complicate matters. In British, French, German and American clocks the piece is simply a lever attached to an arbor that has a means of raising the other levers of the striking movement. In longcase and bracket clocks the lifting piece is outside the movement plates, and the lifting of the other levers is achieved by means of a short lever inside the plates which raises the stop lever and the warning (Figs 198a, 198b, 199 to 201). American and German clocks include types that have the lifting piece within the movement plates (Fig 198c), but this makes no difference to their operation. However, these lifting pieces are frequently made from stout wire, and adjustment of their position and the relative positions of the other levers can best be made by using the tool shown in Fig 202.

In all lifting pieces there are only a few points to look for when correcting faults:

1 The pivoting of the lever must not allow the piece to move sideways and miss the pin that initiates operation. It is unusual for the post or the bearing of the lifting piece to wear to this extent naturally, it is likely to be the result of improper repair or the use of a part from another clock. Correct by bushing or making another post.
2 The piece should drop off its operating pin before the rotation of the wheel that carries the pin starts to lower the lifting piece again. This fault will only occur if the piece is too long, which brings us back to the same causes as in (1); levers do not *wear* longer (Fig 203).
3 Too short a piece will fail to raise the other levers of the movement. This will either be the result of wear or a desperate attempt by previous repairers to solve a problem they did not understand. It is surprising how often this sort of thing occurs, but it is almost never the case that removing metal from a clock will 'make it better'. Even dressing pallets will not do this, but simply makes use of the notable tolerance of the anchor recoil to a wide range of inaccuracies.
4 Polish all parts of the lever that bear on pins and other levers, remove all notches due to wear, despite comments in (3).

Stop, warning and count hook

These all interact with one another, and therefore it is simpler to discuss them in the same section. British, French and German clocks make use of straightforward count plates (locking plates), the latter being rotated by a pinion or attachment to the great wheel so that it makes either 1/78 or 1/90 of a turn every time the strike wheel lifts the hammer tail (Fig 204). At the same time the stop wheel or hoop wheel makes one turn, and the warning wheel makes at least one. The importance of the two numbers, 78 and 90, is that the former is the number of strikes made in twelve hours by an hour-striking clock, and the latter the number made by a clock that also strikes the half hours.

The train that supplies these motions is controlled by the count hook. If the hook can fall into a space on the count plate or wheel it allows another lever, the stop, to fall into the space in the hoop wheel or to catch the pin on the stop wheel. Hoop wheel and stop wheel are alternatives (Fig 199). There is a variation on this type of count control which uses pins in the count plate. In this instance the train is stopped when the hook (for want of a better name) cannot fall into the space between pins – ie, when it drops directly onto a pin (Fig 205, a to c). This system is quite rare.

Testing count-plate operation

Here are a few points to check for correct operation:

1 Does the warning piece catch the wheel before the stop wheel is freed or the hoop lever lifts right out of the hoop (Fig 206)? If not, the train will start to strike instead of being caught on the warning. Adjustment of British clocks will almost certainly entail dismantling the train; French clocks often have a friction-fitted stop piece, so that the relationship between the warning and the stop piece releasing the stop wheel can be set and reset quite easily. American clocks, and some German, only need to have the wire of warning or stop piece bent to alter their operation. Whatever adjustment is made, it should be done by small steps; it is very easy to misjudge the effect of altering the interaction of the levers.

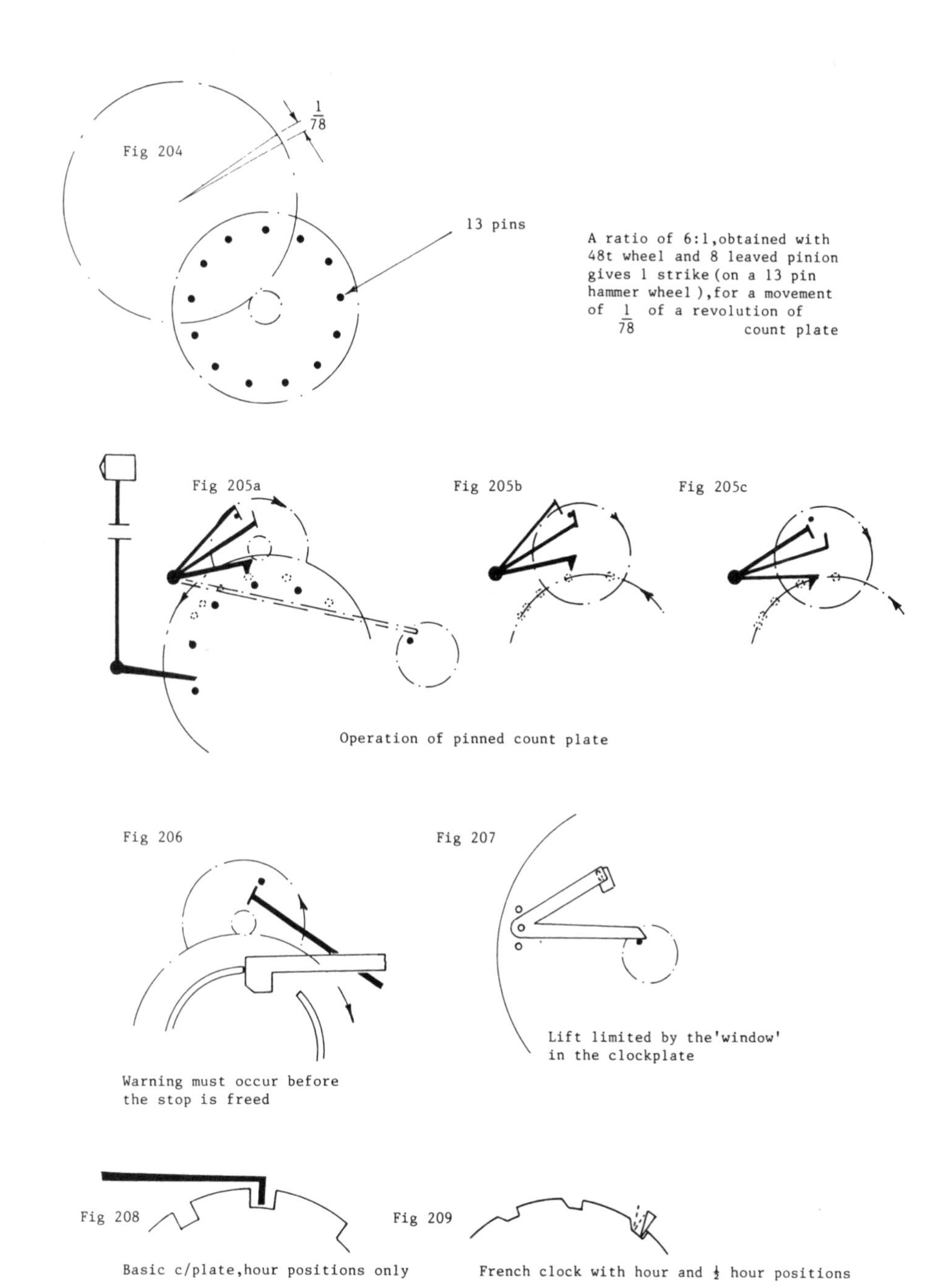

Operation of pinned count plate

Warning must occur before the stop is freed

Basic c/plate,hour positions only

French clock with hour and ½ hour positions

2 Will the warning piece and lift ride to the top of the lifting pin's path without interfering with the clock plates (Fig 207)? Unless the pin shows evidence of having been bent outwards or replaced, do not adjust the pin, attend to the lifting and warning pieces alone.
3 In addition to placing the warning piece ready to catch the warning wheel, the movement of the lifting piece raises the count-plate hook to be clear of its slot and then releases the train by lifting the stop piece away from the hoop or pin that is holding it. It is important that the train does not start moving until the hook is out of the slot, particularly in the type of American clock that has narrow slots with very little room on either side of the hook. Despite its name the hook is not a hook, it is a sensing device; and the only part of it that should touch the count plate or wheel is the end of the turned-over portion.
4 The stop piece and the hook can have their relative positions altered on all clocks by either bending or making use of the friction fit provided by the original maker. When the hook is riding on the unslotted portion of the count plate or wheel, the stop piece must be clear of the hoop or stop pin; in American clocks with a toothed count wheel (Figs 198c, 210), this means the bottom of the tooth spaces. When the hook enters a slot it should enter deeply enough to obtain secure locking on the hoop or stop wheel; it is not necessary for the hook to rest on the bottom of the slot.
5 'Secure locking', in the case of the British hoop-wheel type of clock, will be obtained only if the chamfer on the stop piece drops right through the gap in the hoop so that the face that is radial to the hoop wheel is contacted firmly by the hoop (Fig 211). You may well find that the junction of the chamfer and the radial face has become rounded over the years, and this will tend to throw the stop piece out of the hoop space when the train is rotating.
6 Clocks that employ a stop wheel with a pin in it, will work reliably if the outside of the pin is level with the edge of the stop piece when they engage. Since the stop piece has two sides to it, this means the minimum engagement that is satisfactory, ie when the pin has just moved over the edge of the stop piece (Fig 212).
7 Stopping on the fly is a very good way of bringing the train to a halt, but for best performance the fly should be a tight fit on its arbor so that the movement is stopped when the hammer is off the lifting pins. If the fly is allowed to turn on its arbor – a necessity in other movements – there will be times when the striking train has to start operation with the hammer partly raised.
8 Many American wall clocks employ a 90-tooth wheel with deep cuts in the spaces between the teeth, either gear teeth or ratchet form, for the sensing of the hours and half hours. As long as the hook falls into the space without touching the sides and pulls out in the same manner, the count will operate properly. However, British and European clocks with half-hour strike normally have the cutaways wide enough for the hook to drop back into the slot after the hour strike (Figs 208, 209, 213). This allows one turn of the stop wheel and one strike of the hammer for the half hour, after which the extra space in the slot is used up; and the next time the mechanism is freed by the warning wheel the hook measures out the strike for the next hour.

Make sure that the hook drops into this slot as near to the beginning of the space as possible without fouling the side, otherwise it is quite possible that there will not be enough left for the 'drop back' when the half hour is required (Fig 213). The hook will probably jam on the far side of the slot, or bounce out and put the strike on to the next hour, instead of coming to a halt on the stop wheel. In this type of clock the slot that marks the end of the 12 o'clock strike is wide enough to give a single strike for half-past 12, 1 o'clock and half-past 1, and has a very slender slice of metal out to the outside diameter to give two strikes at 2 o'clock. This slice is so slender that it is not infrequently damaged, or missing altogether. Replace it as you would a missing tooth in a wheel. Saw down the middle of the scar, file out to the correct width and then solder in a piece of well-beaten brass to give the same height as the other plateaux that demark the other hour strikes (Fig 214).

Re-assembly

Putting the striking train together again after

Fig 210

American count and stop

Fig 211

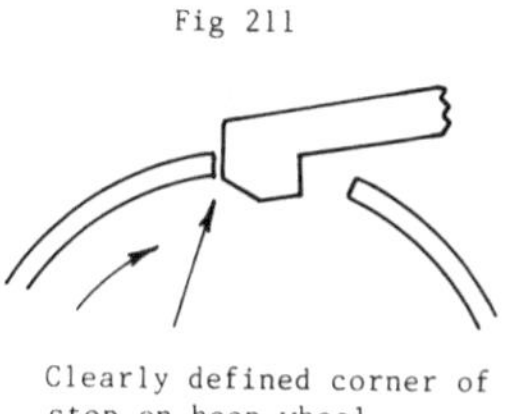

Clearly defined corner of stop on hoop wheel

Fig 212

Position of stop-pin

Fig 213

Position of hook in count plate

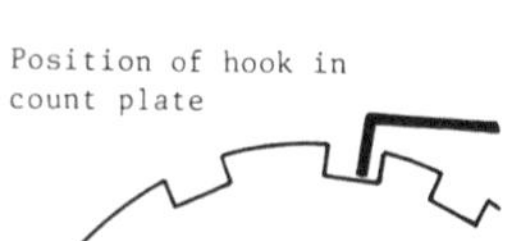

Fig 214

Count plate repair

Fig 215

Stop and hammer positions

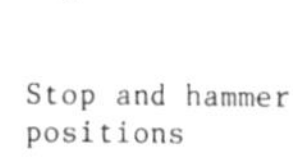

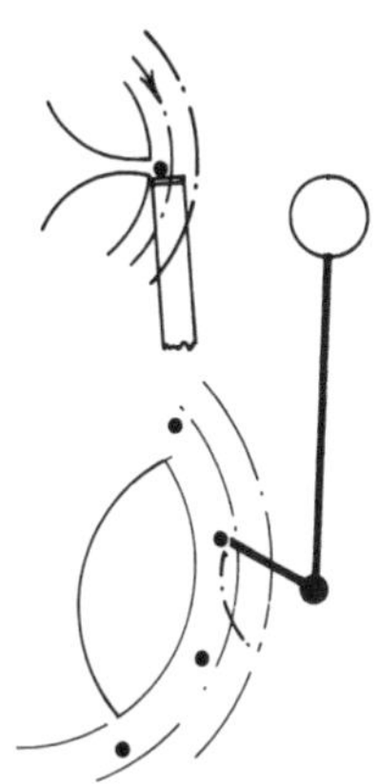

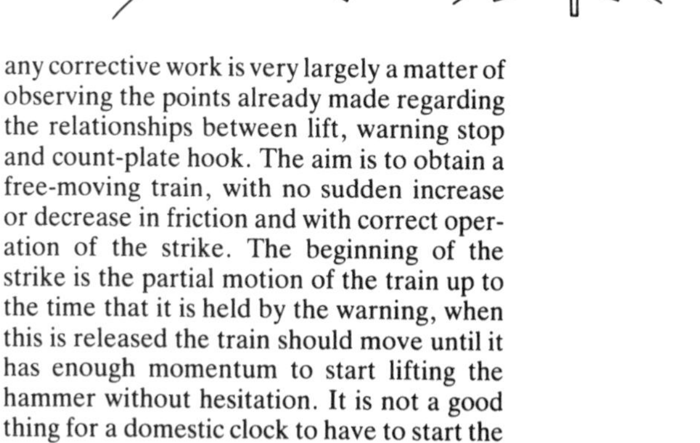

any corrective work is very largely a matter of observing the points already made regarding the relationships between lift, warning stop and count-plate hook. The aim is to obtain a free-moving train, with no sudden increase or decrease in friction and with correct operation of the strike. The beginning of the strike is the partial motion of the train up to the time that it is held by the warning, when this is released the train should move until it has enough momentum to start lifting the hammer without hesitation. It is not a good thing for a domestic clock to have to start the train moving when the hammer is already resting on the hammer lifting pin, but adjustment of the train and the various levers differs between the various types of clock.

BRITISH LONGCASE AND LANTERN

Count plates appear on eight-day and thirty-hour clocks, the former generally having the count plate between the movement plates and the latter having it outside (this, of course ignores the differences that occur in posted clocks with count plates, ie lantern and birdcage clocks). Only the thirty-hour type allows adjustment of the relative positions of the hammer pins and the count-plate hook. It has a pinion driving the count wheel from the outside end of the great-wheel arbor and, when this pinion has a different number of leaves to the number of strike pins on the great wheel, the adjustment is possible. In all other cases the hook adjustment is made by bending the short leg of the hook or its main shank; alterations to the stop piece can only be made by dismantling the clock. Make sure, when assembling the movement, that the tail of the hammer arbor is just coming off a hammer pin when the stop is about to enter the hoop gap or engage the stop pin (Fig 215). You do this by laying the components of the train on the back or the front plate, and being careful not to disturb the meshing of wheel and pinions when the other plate is fitted, but it is not an easy task. It can help to use Sellotape or Blu-tack to hold the wheels and pinions in the required mesh whilst assembly takes place, but beware of leaving any stickiness or bits on the teeth. Whenever possible I leave the hammer spring off the movement until all is fitted together, otherwise it tends to get in the way.

FRENCH CLOCKS

The cheaper French movements have the same problem as the British movement – the wheels and pinions must be arranged in their correct position before assembly. Very often you will find that a wheel tooth is marked with a dot; this tooth should mesh with a leaf on the meshing pinion that has been slightly chamfered or had a dot placed on its face. Again, it helps to stick wheel and pinion together so long as the parts are left clean and not sticky afterwards.

Fig 216

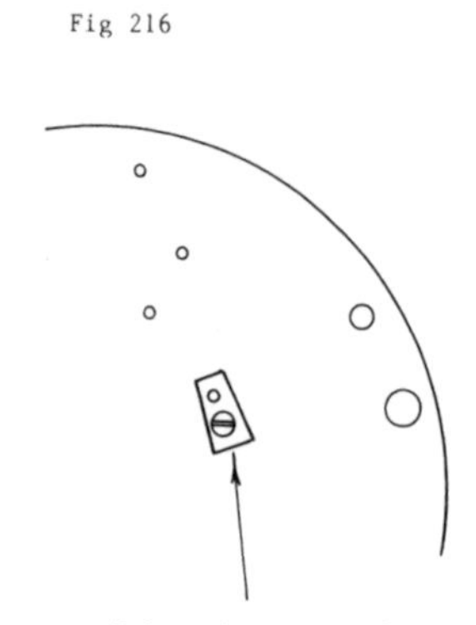

False piece or cock

More expensively made French clocks have false pieces on the front and back plates that carry the pivot holes for the hammer wheel and one or two other pivots (Fig 216). This enables the repairer to change the position of the hammer wheel in relation to the pinion of the stop wheel. It is very likely that such a movement will be partly wound up when the striking fault is discovered; in which case *never* remove the false piece until the barrel has been wound down, as this would have much the same effect as taking the back plate off when fully wound. A last, despairing method of adjustment that I have seen and deprecate, is the drilling of a new hole for the stop pin in the required position. This was done while the movement was assembled.

GERMAN AND AMERICAN CLOCKS

Although both countries produced some clocks that followed French or British methods, they also made much greater use of flexible parts for the strike mechanism. Nevertheless, bending the wires will only apply to a relatively minor adjustment, the wheels and pinions should be set out as near as possible to their correct position when assembling. Since the plates are usually thin, the arbors flop over to a greater extent than would happen in a French or British clock and this adds to the difficulty. A tool for bending the wires was shown in Fig 207.

HAMMER TAILS

Possibly as a result of attempts to adjust the strike by unorthodox means, the hammer tails of French clocks, in particular, are often not the originals, and they may be too short or too long. The too short hammer tail just does not strike, so will be noticed quite easily, but the over-long one is less obvious. It will, however, result in the hammer tail dropping off one hammer pin and directly onto the next (Fig 217); this is not good for the pins or the hammer, and it prevents the train being set up so that it has a free run before making the strike. Much time can be wasted moving the wheels round in relation to the pinion when, in fact, an over-long hammer tail makes it impossible to obtain a proper solution.

Fig 217

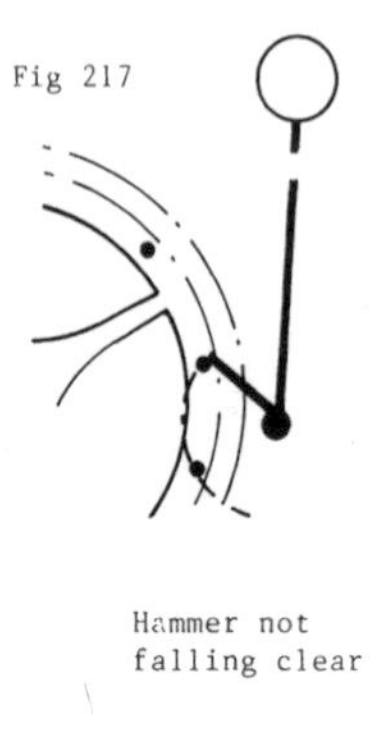

Hammer not falling clear

RACK STRIKING

This is the alternative to count-plate striking. It permits the owner of the clock to move the hands on without allowing each hour to be fully struck; it can also be made to repeat. The control for the amount of striking is a cam that is tied directly or indirectly to the hour wheel; the cam does not advance as a result of the strike being operated, but only as the hour hand rotates around the dial of the clock. If the hands indicate that the hour is 5 o'clock, well clear of the time that the hour is advanced, the clock can be operated as often as you like and only make five strikes, unless there is a failure in the system. Fig 218 shows the main features.

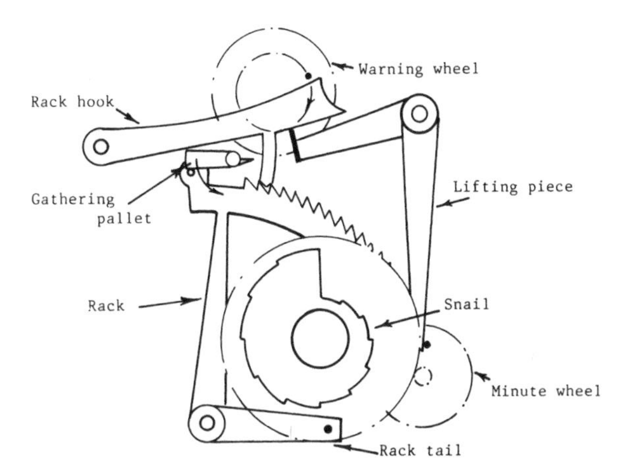

Fig 218

Stopped condition with the rack hook about to be lifted

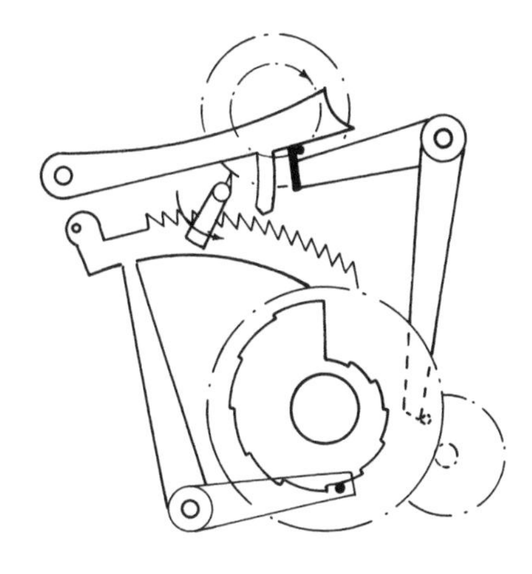

Warning has taken place and the rack has been allowed to expose six teeth to the gathering pallet, the train is held by the warning wheel and the warning arm of the lifting piece

The warning wheel is released and the gathering pallet gathers the rack teeth, one at a time,until the rack and gathering pallet are in the stopped condition once more

The cam is called a snail, and it can either be mounted on the hour pipe or set off to one side on its own post and fastened to a twelve-pointed star wheel. In the first instance the time of changeover from one hour to the next is not clearly defined because it simply depends on the hour wheel slowly turning as time passes. The snail is stepped in most clocks and each step adds one strike to the action. At the junction between one step and the next there is an area where the clock may read the number of strikes required from either of the two steps (Fig 219).

The snail that is attached to a star wheel is advanced by the movement of the hour wheel also, but by means of a peg that engages the teeth of the star wheel. The latter begins to move, and the step on the snail maintains the required number of strikes until a point is reached when a spring-loaded 'jumper' that presses on the teeth flicks the star wheel forward so that the next step on the snail is presented (Fig 220).

Early rack striking utilised a rack and pinion. The pinion was mounted on the end of the pin wheel and the rack acted as a transfer motion between pin wheel and snail; there was not necessarily a one to one relationship between the number of teeth on the rack and the number of hours to be struck. The more familiar rack and gathering pallet, which has one tooth 'gathered' for each strike of the bell or gong, came into general use in clocks made in Britain in the last quarter of the eighteenth century, having grown in popularity over the previous fifty years. Only the rack and gathering pallet are dealt with here, since the early type appears on clocks that are outside the scope of a book on repair and almost certainly calls for high-quality restoration work. The main parts of the pallet system are: rack, snail, rack hook or pawl, lift and warning, gathering pallets. But before going on to the main faults in turn, the system ought to be tested for 'foreigners'.

British longcase clocks that have been left in an attic, or come from a junk shop or unsophisticated antique shop, have often had their movements removed from the case and the dials taken off. The result is that there has been a tendency for bits to fall off, and the unscrupulous seller keeps a box of 'spares' to replace missing parts from. Since the longcase is not made with a standard movement, the chances of these spares matching each other are fairly remote. They may fit, but they are unlikely to give anything like proper operation. It is best to decide whether there is a foreigner in the movement or not and, if there is, to either throw it out or use it as a blank to make a new part. It usually takes longer to try and treat it as a repair.

Persuade the clock to strike the hours of 11, 12 and 1 o'clock; if the train is faulty, take everything out except the wheel with the gathering pallet and turn this by hand. If the front-work, (rack, rack hook, lift etc) has matching parts, the hours will be indicated by the correct number of teeth being gathered for the corresponding positions of the snail. If it fails on any of those three strikes, suspect the use of a spare part. A very mangled rack tail will also give a false strike, and you will have to make up your mind as to which part is the intruder, or whether the tail is at fault, by judging the appearance of the various parts of the front-work. Since these clocks had a great deal of individuality built into them, there is very often a definite, recognisable style; look for a part that does not have similar embellishments or lines to the others.

The rack

Simple inspection will discover teeth that are bent or damaged, you should also check that the pitch of the teeth (the distance between the top of one tooth and its neighbour) is constant along the arc of the rack. This can be done with a vernier calliper, or by filing a small piece of metal to the shape shown in Fig 221. The rack consists, essentially, of a lever with arms that have to be in precise ratio to each other, a pin to register on the plateaux of the snail and, of course, the teeth. The ratio of the two arms of the lever is the same as that of the increment of the snail and the pitch of the teeth. In other words the arm lengths are arranged so that the amount that the pin is lifted at each step between hours, is exactly right for moving the other end of the rack the distance between one tooth-top and the next. If this is not so, the system can be corrected by several methods – the tooth-pitch can be altered, the length of the arm that holds the pin (the rack tail) and the steps of the snail can all be changed. Apart from the undesirability of modifying original parts, it is clear that carrying out alterations

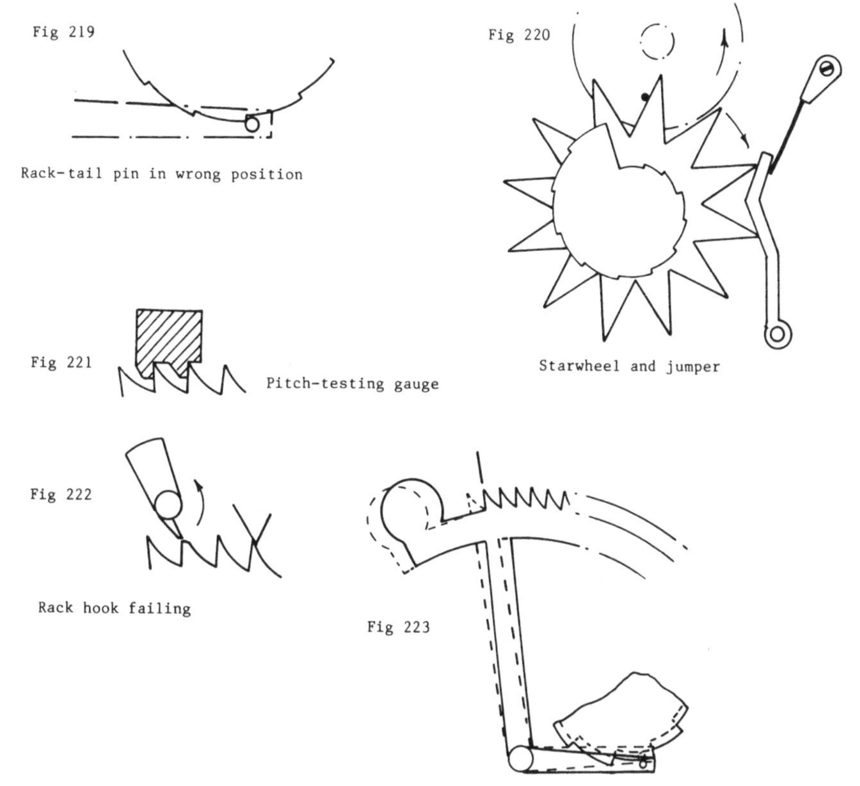
Fig 219

Rack-tail pin in wrong position

Fig 220

Starwheel and jumper

Fig 221

Pitch-testing gauge

Fig 222

Rack hook failing

Fig 223

Testing rack against a mark and checking snail

at random on the snail and rack will do nothing except confuse matters. In all that follows it is assumed that you have viewed the movement and decided to make adjustments on one part at a time, and carried that through until you are certain that the part is correct for the clock, before touching anything else.

The pitch of bent teeth is corrected by using smooth-jawed pliers and bending the teeth until the radial face is a straight line again. If any tooth is shorter than the others it must be removed and a new tooth inserted or, if the shortness is slight, the other teeth shortened to match. You can judge whether a tooth is too short or not by trying the gathering pallet on that part of the rack, and seeing whether it can throw the tooth far enough to advance the rack by one tooth space (Fig 222). Carry out this test with one finger holding the rack up to the pressure of the gathering pallet; the rack hook may have to be altered to operate correctly and using it at this time to position the rack teeth will confuse matters.

When the teeth are all even it is time to match the rack against the snail. Begin at the largest radius of the snail, ie 1 o'clock. When the pin is resting on this plateau, one of the teeth of the rack should be sighted against some mark on the plate; any mark will do but it should be amongst the first two or three teeth of the rack so that all the rest can be checked as the snail is turned (Fig 223). Move the snail through 2 o'clock, 3 o'clock and so on, making sure that consecutive rack teeth line up with the mark accurately. Since we know that the rack teeth have constant pitch, any variation between a rack tooth and the mark has to be either a fault of the ratio between the two arms of the rack, or of the increments on the snail. An uneven variation

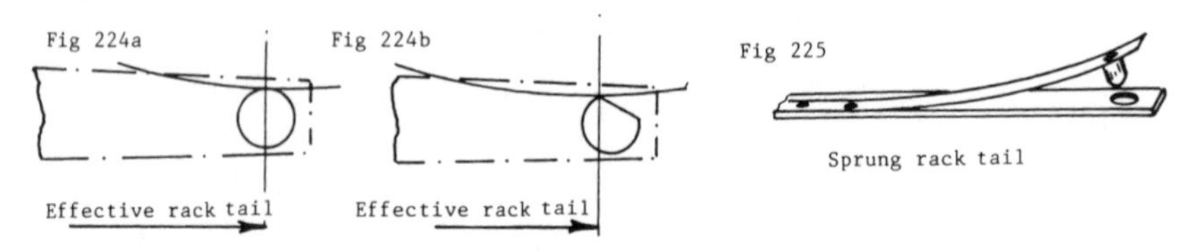

Adjusting the length of the rack tail

(gaining on one tooth, losing on another) will require you to remake at least part of the snail. A steady gain or loss can be put right by attending to the rack tail or the snail, the former being the easier job; but if the rack is in good condition and the snail looks as if it is a later addition, then a new snail is a better answer from the point of view of retaining the original clock.

Small alterations to the radius of the pin in the rack tail can be made by carefully filing a flat on the pin (Fig 224), but anything over 1.27mm (0.05in) calls for re-locating the pin, which in turn will require you to make another rack tail. The latter was made from flexible, hard brass about 0.25mm to 0.38mm (0.01in to 0.015in) thick on most British clocks prior to the twentieth century; German and American clocks mostly followed the same pattern, but French clocks were usually fitted with rack tails that will not flex. The advent of the factory-made clock of the late nineteenth century onwards, saw rack tails made in two parts – a stiff support with a hole in it where the pin was required, and a flexible flat spring that carried the pin. The pin fitted through the hole and was prevented from distorting when hitting the outside of the snail by the stiff support, but was driven back into the hole by a ramp on the snail if the rack failed to lift up to its fully gathered position (Fig 225). This does not originate with factory-made clocks, but was not common before then. The construction of a new rack tail for this type follows the same method as for the earlier ones but, after establishing the position for the pin, the hole is opened out to accept a pin mounted on the flat spring.

Remove the old rack tail, and make sure that the pipe that forms the pivot of the rack is firmly riveted to the part of the rack that carries the teeth. Choose the correct material for the particular type of rack tail, making it twice as wide at the outer end as is necessary and a little longer. Drill a hole to fit tightly on to the pipe and, if the pipe was damaged when the old part was removed, re-turn the seating for the tail. Fit the new tail and make sure that although it can be twisted around the axis of the pipe, it does not do it easily.

Make a guess as to the position of the new pin and mark it on the edge of the tail with a pencil. You must now check that this is right or wrong by going through all the hours, lining up the mark on the corresponding part of the snail and seeing that the teeth neither gain nor lose on the mark made on the clock plate. If the points of the teeth are not moving far enough at each hour, the distance of the pencil mark on the tail from the centre of the pipe must be decreased; if the teeth are moving too far at each hour the length must be increased. As soon as an acceptable position has been found by this means, drill a small hole 1.27mm (0.05in) diameter close to one edge of the tail and on the pencil mark (Fig 226). Drive a taper pin into this hole and go through the testing routine again. There should be no need to make any great amount of correction now, but slight movements can be made and a new hole placed on the centre-line of the tail for the pin (or for the pin to pass through in a modern clock). File away the edge of the tail that bears the test-hole.

The pin usually has a diameter of between 2.01mm and 3.05mm (0.08 and 0.12in) and, as you can see from Fig 224, the actual contact point with the snail can be moved inwards or outwards by filing flats on it. Note that there is a ramp on the base of the pin.

COMPLETION OF RACK SETTING

The rack teeth are now correct and so is the ratio of the two arms of the rack, but it is necessary to position the two arms so that the right tooth is placed for gathering at each hour. Place the rack in the stopped position (Fig 227). British, most German and American rack-striking clocks prior to the twentieth century have a tail on the rack that stops the striking train by locking against a pin on the rack. French clocks stop on the stop wheel, and the rack is positioned for first strike by the rack hook (Fig 228).

With the rack held by the gathering-pallet tail or the rack hook, the tooth nearest to the centre of the gathering pallet counts as zero. Move the rack until tooth number 1 is opposite the centre of the gathering pallet. The tooth need not quite reach the centre, but it must not go past centre or you will find it difficult to gather a full pitch and risk not catching the top of the next tooth to be gathered. Hold the rack in this position and move the tail until the pin comes into contact with the plateau for 1 o'clock. Gently test this for the rack and snail positions of 11 and 12 o'clock and, if all is well, take the rack very carefully off its post. The tail can now be fastened permanently by heating and running a very small fillet of soft solder around the joint between the pipe and the tail (Fig 229). It helps to have tinned both these parts before assembling. A secure fillet will show just a thread of bright metal when the solder is molten, more will look ugly and is quite unnecessary. After soldering, test the rack through all the hours in case anything has shifted in the process.

Snail

The snail, as already indicated, is a cam that progressively moves the tail of the rack a constant amount (the increment) for each succeeding hour. In most cases this is done by making plateaux – one for each twelfth of the circular surface; and each plateau has a constant radius for the extent of that twelfth (Fig 230).

Some cheaper nineteenth-century French clocks have a snail with no steps, but a smooth curve. This has the advantage of

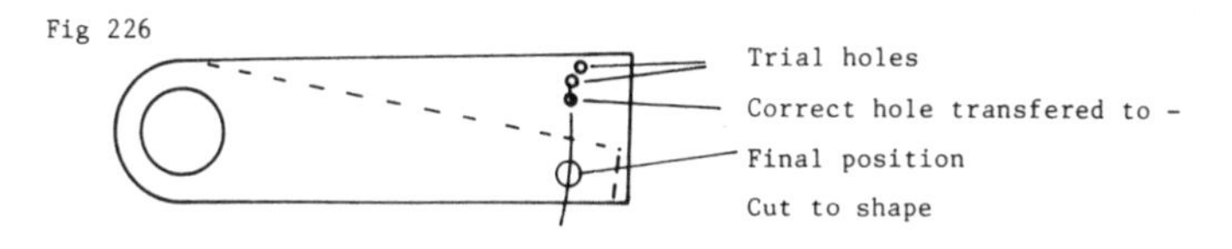

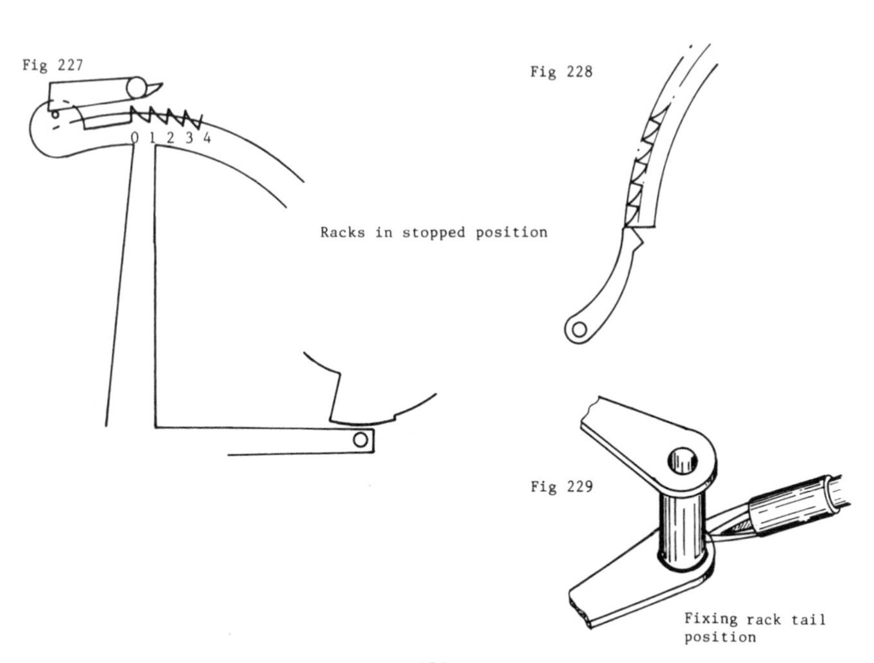

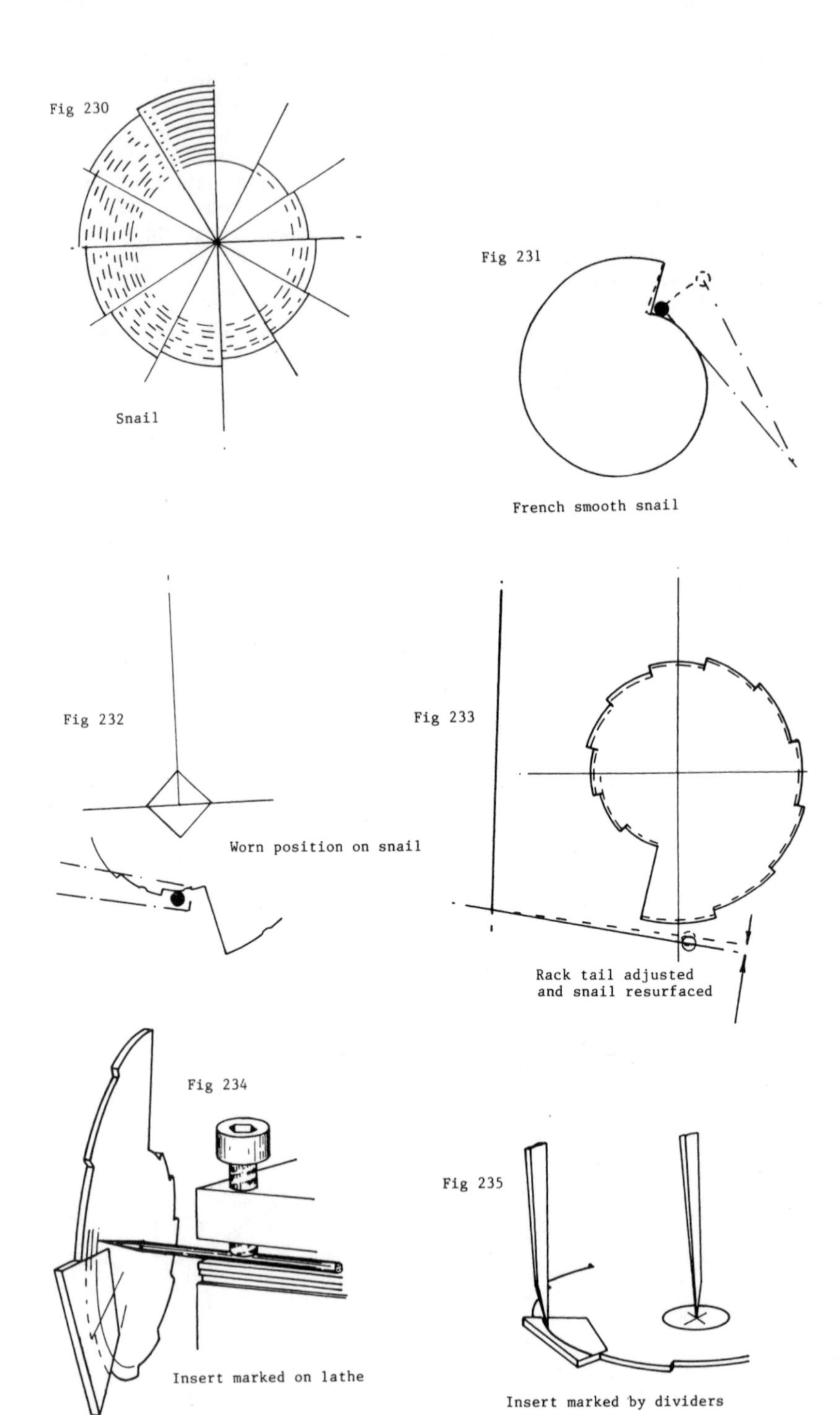
Fig 230
Snail
Fig 231
French smooth snail
Fig 232
Worn position on snail
Fig 233
Rack tail adjusted
and snail resurfaced
Fig 234
Insert marked on lathe
Fig 235
Insert marked by dividers

making it unnecessary for tight control of the angle between the rack and its tail during the original production, but can make it difficult to determine a good position for the rack-tail pin when a notch has been worn in the once-smooth surface. It can be helpful to file the large step between 12 o'clock and 1 o'clock so that it is no longer radial, but gives more space for the rack tail to swing in and out of the snail, extending the position at which it can register twelve strikes reliably (Fig 231).

Most snails will show only a slight amount of wear, in the form of a hollow in one part of the plateau. It is not necessary to make this hollow level again by soldering or replacing with a piece of brass, the problem can frequently be solved by making sure that the rack falls onto another part of the plateau. If the hour hand is fitted onto a square, or pinned to the pipe in a way that cannot be changed (Fig 232), the snail will only expose an unworn part to the fall of the rack tail if the hour hand is shifted as well, and this will mean that it does not precisely point at the hour when the strike takes place. If the error is too glaring to be acceptable, the snail can be removed from the hour pipe and each of the worn plateaux filed until the hollows disappear. Remove the same amount from each one – a vernier can be used to measure the distance from the inside of the pipe to the plateau being filed – and keep each one a true part of a circle. When all is done, the arms of the rack will need to have the angle between them opened up slightly, so that the right teeth of the rack are offered to the gathering pallet (Fig 233).

If the damage is confined to one or two plateaux, it may well be easier to cut back the worn faces and solder in a piece of brass strip. Deciding on the position on the faces after filing will be difficult unless a little exercise in marking-out is undertaken. If you have a lathe, hold the pipe in the chuck and hold a scriber in the tool-post. Choose plateaux on either side of the worn part and line the scriber up with the radius of the smaller one. Turn the chuck by hand and carry this small radius over to the larger radius on the other side of the worn part, make a mark with it (Fig 234). Use the dial on the cross-slide to measure the distance between this marked radius and the larger one, not forgetting to allow for the back-lash of the lead-screw, and divide this figure by the number of plateaux to be corrected. Position the scriber using the slide and this figure for the increment, and mark each of the plateaux that have new metal soldered on.

Lack of a lathe makes this a little more difficult, but well within the reach of average hand-tool skills. Fill the end of the pipe with a piece of lead sheet. Find the centre using dividers and mark this centre strongly – it will probably not stand the use of a centre-punch. Use the dividers again, and carry the smaller radius of the two that flank the worn plateaux onto the larger part. Measure the distance on a good steel rule or with a vernier calliper, find the increment by dividing as before. Use dividers to mark the face of the snail with the new radii (Fig 235).

Some snails are supported on a separate post and moved around each hour by a star wheel and operating peg. These are often friction fitted, allowing movement of the snail. When they are not friction fitted, they are held with two or three screws and are easy to remove. You will find that damage of the sort we have just discussed rarely occurs on lighter clocks, such as French 'marble clocks' and carriage clocks, and is more likely to be the result of bad treatment than of wear.

Rack hook (pawl)

The task of the hook is to prevent the rack falling back after a tooth has been gathered, but it also serves to position the teeth correctly so that the gathering pallet picks up a tooth cleanly and leaves it a full pitch further on. As you can see from Fig 236, the hook has two angles that rest between the rack teeth so that, as the tooth passes under the blunt point of the hook, the weight or spring of the rack hook moves the teeth in the same direction that the gathering pallet is taking them. Fig 237 explains the effect that filing the angles has on the position of the rack, when the gathering is complete. A faulty rack hook is a common reason for the clock to strike thirteen or more times, or even to strike continuously.

On most antique British clocks the hook is gravity operated; check the post and the freedom of the hook pipe on it. Corrosion and old oil will often make the operation sticky, and wear on the point of the hook and the point of the gathering pallet will affect the correct amount of movement of each tooth. Wear on the tooth of a gathering pallet can be compensated for quite frequently by an adjustment of the angles of the hook – removing metal from the left-hand angle will move the rack to the right, and removal from the right-hand angle will move it to the left. Of course, the point will also be moved; and you must be careful that it is not moved so far as to no longer lift over the rack tooth.

Most French 'pendule de Paris' and carriage clocks have the hook mounted on the stop lever that engages the stop pin. It is spring assisted and, although usually made of brass, I have never yet had to modify a French hook. Most of the faults on the gathering of rack teeth have been due to chipped, cracked or lost gathering pallets, other faults lay in damage to the teeth or curvature of the rack itself.

Lift and warning pieces

The comments that were made on lifting pieces and the warning lever in the paragraphs on count-plate striking (page 112), apply also to the rack strike.

Gathering pallets

There are three main types of pallet – British, French and modern. The first two appear on antique clocks and the third on late-nineteenth and twentieth-century ones (Fig 238 a, b, c).

BRITISH PATTERN

This type carries a tail and brings the striking train to a halt by the interference of this tail with a peg on the rack (Fig 238a). Quite a lot of shock is absorbed by the extension that carries the pallet when the train is stopped but, surprisingly, the extension is rarely broken off by this. Clockmakers have had different approaches to the proper functioning of this gathering pallet over the years, and it seems best to look at the different methods of making the item.

Tooth shape: The two forms of tooth are shown in Fig 238b. One has its leading side made along the centre-line of the pallet and has a sharp tooth. This type will gather quite small teeth, but has a tendency to become chipped.

The second form has a rounded tooth and is evenly distributed on either side of the centre-line. This is harder wearing, but cannot be used on fine-toothed racks without the curve at the end of the tooth making a clean pick-up of the rack quite difficult.

Tail position: Some clocks have the peg on the rack facing outwards from the clock and some have it facing inwards at the frontplate of the movement; consequently the tail of the pallet will pass in front of, or behind, the rack. From the repairer's point of view, it makes no difference at all, except that the pallet has to be put on the right way round.

Engagement of tail: The original placing of the peg on the rack and of the tail of the pallet that engages with it, varies. Some tails have a curve on the underside, some are flat (Fig 239). The relative positions of the gathering-arbor centre, rack peg and shape of tail, can result in the pallet tail wedging against the peg so that it tends to throw it in the release direction (on British clocks, to the left). This can be so tight a wedging action that the rack hook is clasped by the rack teeth and is difficult to free by normal action of the lift. Alternatively it can have a flat tail to the pallet, that performs no holding or urging function, it does not grip the hook nor does it assist the rack to move to the left when released. The end of this type of pallet is either a radius struck from the centre of the mounting hole, or a relieving curve. This type must not have any notch worn underneath, otherwise the rack can be held back.

Fastening of gathering pallet: The British type of pallet is mounted on a tapered, square-sectioned extension to the arbor. This may simply be a close fit, or the square may be drilled at the end for a fine taper pin, or the end may be screwed and a nut fitted. In all cases, if the pallet is at all tight when the pin or nut is removed, use two levers to prise it off without bending the extension (Fig 240). The levers can be made from silver steel rather like tyre levers or 'jemmies', then hardened and tempered dark straw. They are invaluable for any removal of tight-fitting parts from arbors (Fig 241).

FRENCH PATTERN (Fig 238c)

The striking train of the typical French clock of the last century is stopped by an arm on the same arbor as the rack hook. The gathering pallet, therefore, only performs the function of gathering the rack teeth; it is mounted on

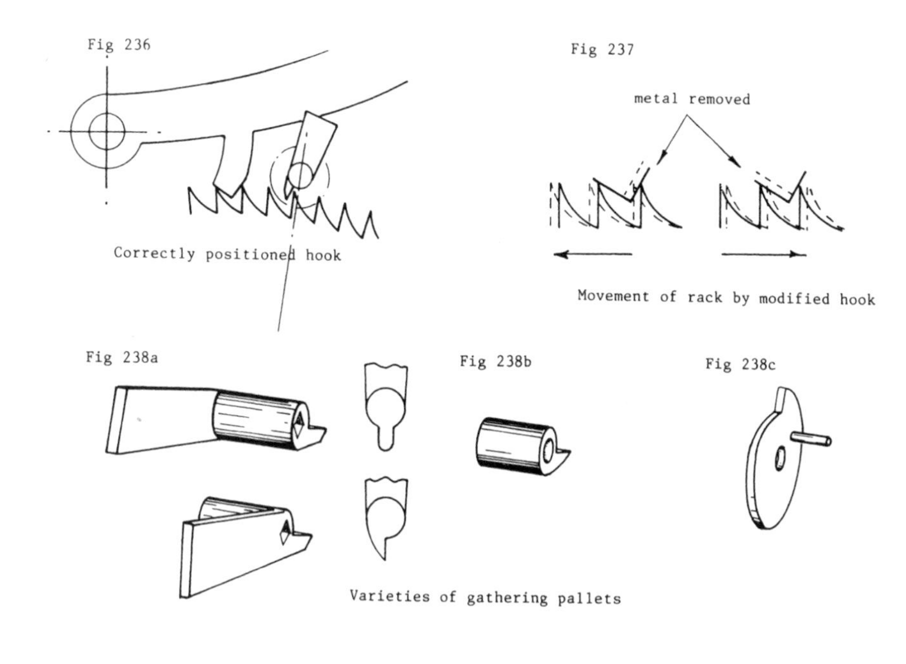
Fig 236
Correctly positioned hook
Fig 237
metal removed
Movement of rack by modified hook
Fig 238a
Fig 238b
Fig 238c
Varieties of gathering pallets

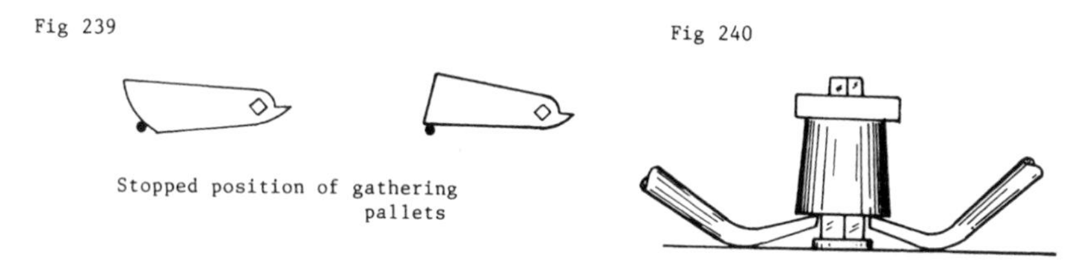
Fig 239
Stopped position of gathering pallets
Fig 240
Using levers to remove pallet

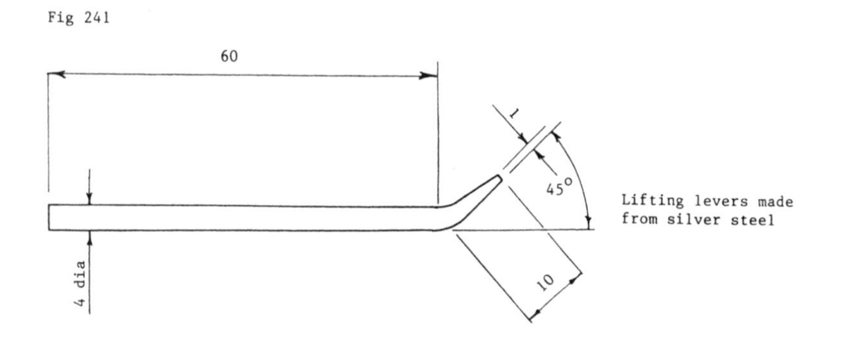
Fig 241
60
1
45°
10
4 dia
Lifting levers made from silver steel

either a square-tapered or a round, extension, and is frequently held by friction alone. The round mounting allows easy setting of the stop, since all that is necessary is to run the train until the hammer trips off the hammer pin, ensure that the stop pin is engaged at this point and then place the gathering pallet onto its extension with its tooth having just come clear of the rack teeth (Fig 242).

On all but grande-sonnerie clocks and some repeaters, where two racks are employed one behind the other, the gathering pallet is a simple tooth, rounded at the back.

MODERN PATTERN (Fig 238d)

This is a cam and pin type. The cam governs the positioning of the stop piece and the pin – stuck into the face of the cam – is set out from the centre of the cam so that when it rotates it gathers the rack teeth (Fig 243). This too is often mounted on a round extension, and will very likely be found to be a tight fit because it was most probably pressed on by machine.

A similar, earlier, gathering pallet is used on some American count-wheel clocks. The arbor that carries the stop wheel also carries a single pin protruding from the end of the pinion (sometimes a lantern pinion with one long trundle), and this gathers one tooth at a time of a circular rack made onto the count plate (Fig 244).

There is no good reason why any work should need to be done on the cam; most of these movements are less than fifty years old and the amount of friction on the cam's acting edge is negligible. The pins, however, are subject to damage when a rack refuses to move or is positioned wrongly by the hook. Although hard taper pins are available, they are not always easily located, and I find that using the lathe to stone a starting taper onto a piece of pivot steel answers very well. It does not have to be a lathe, an electric drilling machine will do just as well.

MAKING PALLETS FOR BRITISH AND FRENCH PATTERNS

Both types of pallet are relatively small and the task of making them will be made easy or hard by the sequence of operations that you adopt. In both cases, as in any small jobs leave the work attached to the material that you are making it from for as long as possible. Carry out the most difficult work first, or that which is most likely to have a false start. There is nothing worse than leaving the most awkward task until the last and then, when the job is all but finished, making a slip and having to scrap all previous work.

British pattern: Choose a stick of high-carbon steel with a cross-section suitable for the dimensions of the pallet. Drill the hole out for mounting on the arbor extension, remembering that the drilled hole must be no larger than the distance across the flats of the eventual square hole.

Use a square escapement file, about 50mm (2in) long, to file out the square, trying it on the extension until it fits. This is not something that can be readily described in a book, you must practise and obtain the practical advice of a skilled acquaintance. The fit should be good, with no wobble, because any looseness will accelerate the wear caused by the shock of stopping the train.

Now file the tooth to the form most suitable for your clock; that in Fig 245 is the one most frequently used by the old makers. Assemble the rack and hook to the front plate and insert the gathering arbor into the two movement plates; clear away as much of the rest of the front-work as possible or practical. The partly finished pallet can be mounted on the extension, and the length of the new tooth checked against the rack teeth. Do not forget that, since the work is still on the rest of the stick of steel, you have a lot of leverage over the extension; it would be very easy to sneeze and snap the extension off.

As you try the pallet against the rack teeth, you must ensure that it only sweeps into the space of the tooth on the entering side of the extension and does not crump down on top of the next tooth-tip. The pallet should just clear the next tooth-tip, and move the one initially under the extension far enough along for the hook to hold it in its final position (Fig 236). Most old makers pushed the rack further than necessary and let the hook bring it back again; this looks untidy, but if you have any doubts about your ability to be accurate it is the safest way.

As soon as the pallet is the correct length, remove it from the extension, place the rest of the stick in the jaws of a vice and, using a file and junior hacksaw, shape the tail of the pallet to its final form (Fig 245). When you

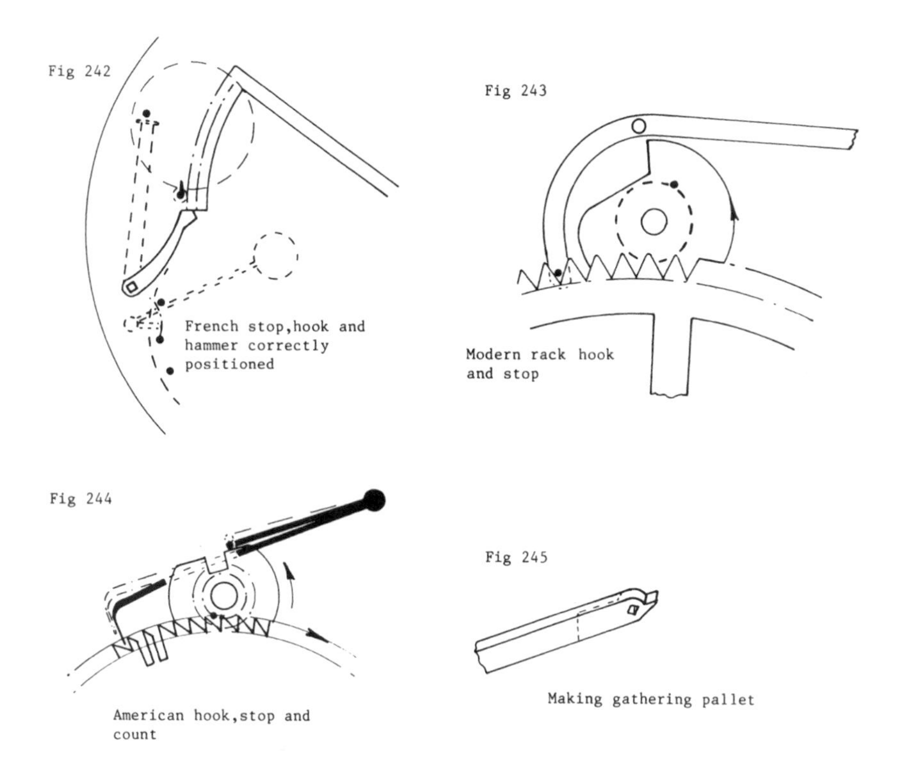

are satisfied, partly cut through the waste so that the new part can readily be snapped off the stick of steel; bring the part to bright red, quench in oil, and follow this with blue temper. When you have done this, the pallet can be broken off the rest of the metal and the end of the tail finished. Note that the length of the tail must be sufficient to stop the pallet when the hook is down and the rack is at its starting position.

French pattern: Again the work should be kept on the stick as long as possible. Make the tooth of the pallet as before, partly cut it off, harden and temper. Snap off and polish the piece.

Alignment of front-work

Front-work is largely made up of levers pivoting around posts. The levers are generally made from flat materials and, so that they do not wobble in the plane of their action, they are mounted on a short tube, or pipe, to give stability. Quite evidently this stability will suffer over the years as the posts and pipes wear, and accuracy of action will be affected by the movement of posts as various owners or repairers handle the clock. Make sure that the allowed movement of each part of the front-work is not distorted, that pieces that should clear each other do so and cannot be made to clash without using force, and that those parts that are supposed to register on each other do so without fail. All points of contact should be un-notched and smooth; re-surfacing these contact points with fine emery paper and crocus paper, backed with a piece of flat metal, is good practice whenever the front-work is removed. Finish with a very light smear of oil.

One last and very important thing; when replacing the front-work make sure that all pins and retaining washers are securely in place. It is quite common to use a thin washer on the post of the compound gear (minute wheel) that serves to restrain the hour wheel by presenting a flat surface to the sides of the gear teeth. Cut the pins so that they are no longer than necessary to do their job.

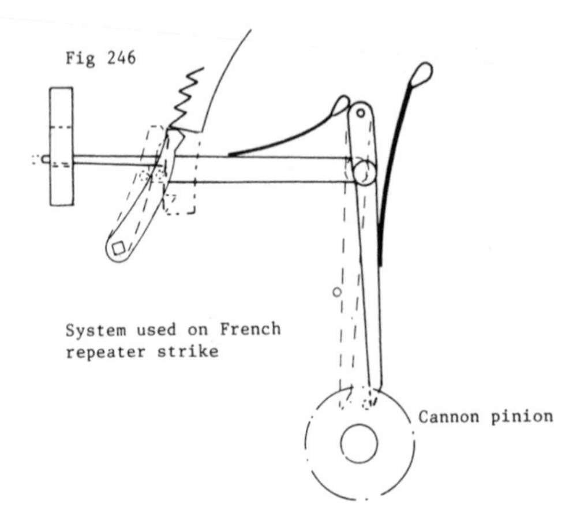

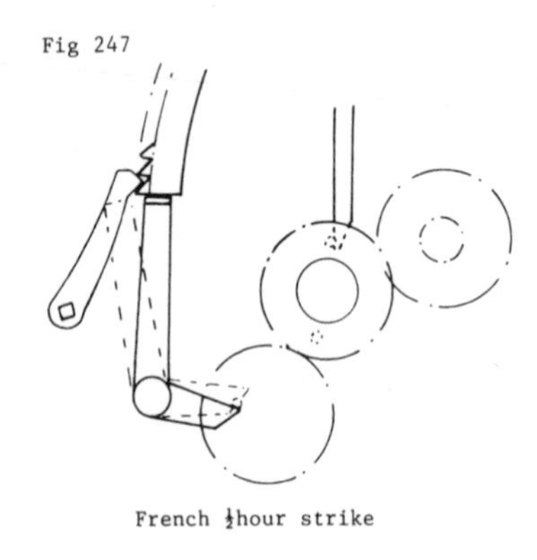

REPEATER STRIKE

This type of carriage-clock movement has no warning, because the warning period would prevent the use of the repeater function for three or four minutes before each normal striking point – hours, half-hours and quarters, according to design. When the warning has operated the train cannot move until the release point is reached.

The need for warning is eliminated by having the lifting pin operate a trigger that is pulled back against spring pressure as striking time approaches, and then slipped off the pin that was pulling it back. The trigger flies forward to hit the rack hook, the hook is knocked backwards and its attached stop lever releases the train, and simultaneously the rack falls so that its tail rests on the snail before the train starts to turn (Fig 246). Points to observe are:

1 The trigger arm must be absolutely free in its guide.
2 There should be little or no oil on these parts.
3 The spring must be strong enough – it gets bent easily – to regularly throw the hook backwards and release the rack.

Repeater carriage clocks have the snail mounted on a star wheel, and the half-hour strike is obtained by moving a stop beneath the rack so that it can only drop one tooth. The pins on the cannon pinion that operate the trigger are of equal length and radius from the centre, and the choice of striking hours or half-hours lies solely with the wheel that moves the stop beneath the rack. It is a nice positive action, and rarely at fault (Fig 247).

CHIMING WORK

There are very many methods of governing the chimes of a clock, so to simplify matters I am going to assume that any chiming clock earlier than late Victorian will be attended to in the workshop of a professional clockmaker or an amateur with a great deal of experience. This leaves us with the typical mantelshelf Westminster chime, produced in Britain, Germany and the USA. Here too there are many differences between the millions of movements manufactured, but the basic logic of the system is fairly constant.

Movement with no automatic correction

The first movement to be described is the type that has to be synchronised with the time shown by the owner, there are no automatic devices to correct a chime and strike that is out of mesh with the dial and hands (Fig 248). The sequence of events is as follows:

1 A pin on the motion work, or an attached lever, raises a warning flag and releases the train; at the correct time the warning is released (Fig 249).
2 The train rotates a pin barrel beneath the chiming hammers at the same time as a small four-notched or four-pinned count plate. The lever that carried out the chime warning rides on the plateaux between these notches, and in this position neither

Chiming arrangement

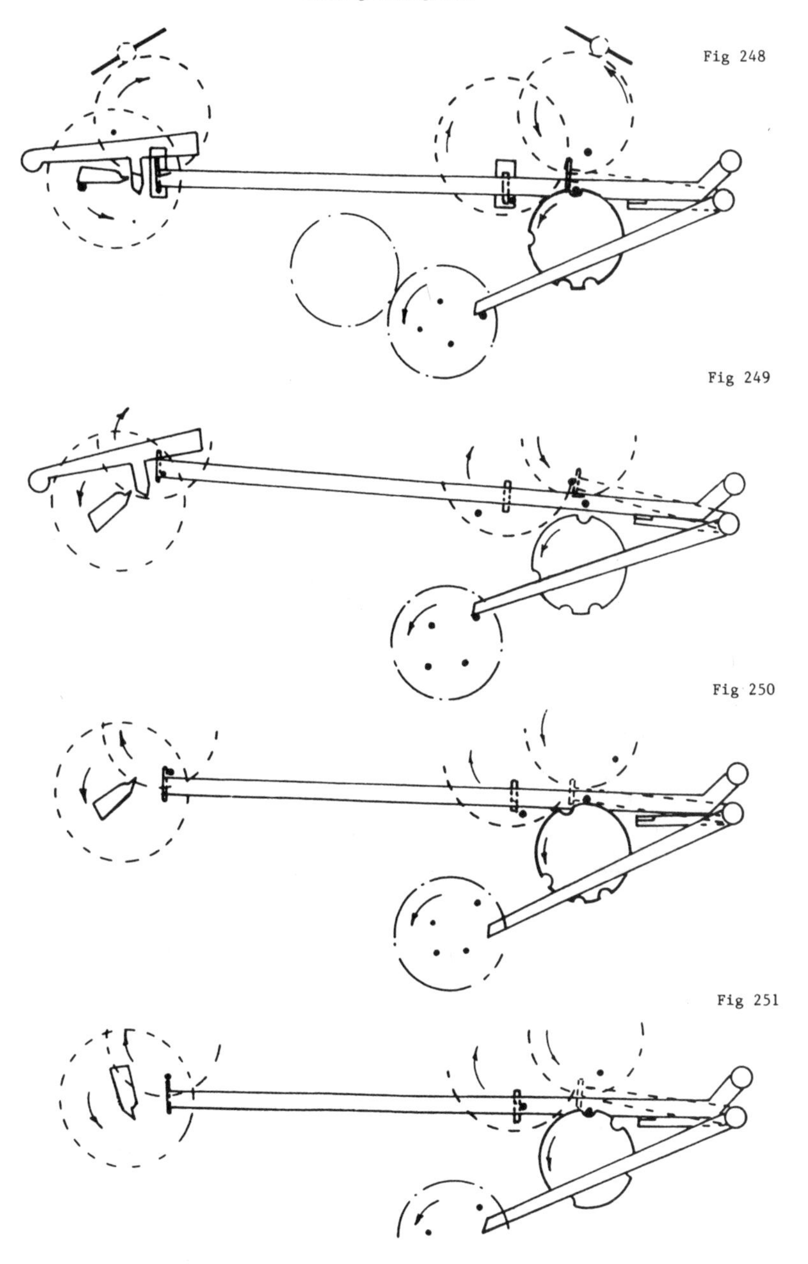
Fig 248

Fig 249

Fig 250

Fig 251

the flag that is intended to arrest the warning pin, nor the detent that will be used to stop the train, is in operation (Fig 250). One of the notches on the count plate arrives beneath the warning lever which then drops, allowing the detent to stop the chiming train (Fig 251). This sequence of events is true for the three quarters but, during the chime that precedes the hour strike, a pin or lobe on the count plate gives an extra lift to the warning lever. Because of its length – it reaches right across the clock – this lever is able to warn the strike train, either directly by means of another flag at its extreme end or by raising a separate warning lever.

4 The warning is held until the count-plate notch allows the first warning lever to drop and release the strike warning.
5 On the strike side the rack has fallen to give the correct number of teeth for gathering, and the whole train is held until the chime has finished.
6 The chime warning drops into the hour notch, the strike train is released and the hours are struck.

This pattern of events gives a reliable pattern of chimes followed by striking, the hours always follow the hour chime and not the quarter, for instance. However, there is no control over the synchronisation of chime and time shown. If the minute hand is moved on swiftly, without waiting for each of the chimes to reach completion, the chimes will still follow the pattern but with no relation to the time. The quarter may well be chimed at the half-hour, with consequent hour striking at quarter past the hour.

There is a strike/chime system which ensures that the hour is always struck on the hour, but does not ensure that the chimes are in sequence with the strike. It can be seen in American two-barrelled chiming clocks, where one barrel performs the functions of strike and chime and there is a sliding arbor that disconnects the pin barrel when the strike is required. In this type of chime the pins or cam on the cannon wheel that raise the lift – the pins can be on cannon or minute wheel in any clock – have one lobe or one pin that raises the lifting piece higher than at the quarters (Fig 252); this warns the striking train and it achieves correct striking, but not necessarily correct chiming.

Fig 252

In both cases, chime and strike can be brought back to their proper sequence by either moving the hands on before the chime has finished until hands, chime and strike agree with one another, or by operating the extension to the chime warning lever that is often provided for the clock repairer's use. This usually has a hole in it so that a thread can be attached and the chime released when the movement is cased. This is the proper use of this extension, it is *not* intended for repeat action, nor waking the baby.

Automatically corrected movement

Automatic correction is more satisfactory from the customer's point of view and you will find this arrangement on the greater number of mantel chiming clocks made from the 1920s onwards. Such a movement employs a four-lobed cam on the cannon wheel with a long lobe corresponding to the hour-striking position (Fig 252). The stop wheel is engaged by two detents, usually hook-like. These are set on an arbor side by side, one being fastened to the arbor by a friction-tight collet and the other completely loose. The loose detent has an arm which rides either on the count plate, or on another single-notched plate within the movement mounted on the count-plate arbor (Fig 253). The detent that is fastened to the arbor is raised or lowered, when the arbor is rotated, by an exterior arm operated by the chime warning lever.

The two detents interact by means of a peg on the friction-tight one that passes through a relatively large hole on the other. The hole may be cut through to the underside of the detent. Thus the loose detent can lift without affecting the friction-tight one, but when that one lifts the loose detent also rises.

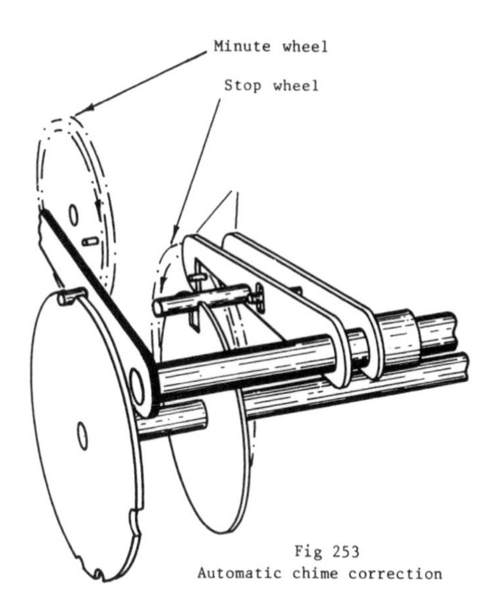

Fig 253
Automatic chime correction

Assuming that the clock is out of sequence at the end of the quarter to the hour chime, the following take place:

1 The loose detent's arm drops into the notch on an interior plate or into a deeper notch on the exterior count plate (deeper that is with respect to the other count plate notches). It hooks onto the stop pin and holds the train.
2 A quarter of an hour later the clock operates the lift for one of the quarters (it is out of sequence remember). The warning on the chime raises the detent on the friction-tight arbor, but not sufficiently to lift the other detent by means of the peg-in-hole interaction. Nothing happens, the train remains detained by the loose detent (Fig 253).
3 This condition exists until the large lobe of the lifting cam, or the corresponding pin with large lift, raises the chime warning lever higher than at the true quarters. The detent lifts higher also, and this time raises the loose detent high enough to release the stop pin. Both detents are now clear of the stop pin, the train revolves and comes to rest on the warning.
4 The chime warning lever drops and chiming takes place followed by the strike.
5 At the end of the strike the count plate or the interior cam no longer permit the arm of the loose detent to fall into a notch, and so this detent does not retain the train. The train continues to be held by the friction-tight detent which is raised every quarter to release the train into the warned state. Any alteration of the hand-to-chime position will bring the correcting detents into operation again.

Since the clock corrects itself, there is no necessity to put the chime and strike into correspondence when the clock is serviced. Do not forget that this will probably mean that for the first few quarters, while the automatic correction is going on, there will be no chime and no strike. The immediate reaction when the clock fails to chime is that there is a fault, and I can recall at least one occasion when this led me to start dismantling the clock again, quite needlessly.

The amount that the two detents are lifted is important. The loose one is unlikely to vary unless someone has strained the position of the arm. The setting of the friction-tight detent is normally by means of a small grubscrew and collar to the exterior lifting piece. Set this screw just tight enough to hold, but allow movement when firmly twisted by hand. The detent should just catch the stop wheel when the long warning lever drops into the count-plate notch. It should also raise the loose detent when the long lobe of the cam on the cannon wheel is operating the lift.

Adjusting the chime tune

The actual sequence of notes is established by the pin barrel, which is often a bank of linked cams; you should have had no reason to alter this. However, the phrases of four notes that make up the chimes have to start at the right time, otherwise the chime will sound very strange. First of all, make sure you know what the chime should sound like at the quarters and at the hour.

1 Move the movement onto the end of the hour chime and strike.
2 Rest a light bar on the top of the movement to engage the fly and prevent unwanted progression of the train operations. You will want the train at this present position, poised to begin the first quarter chime.
3 At the back of the movement there is

usually an exposed gear that drives – through one, two or three idlers – the chiming or pin barrel. Undo the grubscrew that permits this gear to turn without moving the rest of the train and, by turning it manually, take the chimes through their complete sequence.

4 When you are sure that the next phrase will be the quarter-past chime, lock the grubscrew and then test the movement after removing the bar from the top. Any slight adjustment of the chime barrel, forwards or backwards, may be made by simply slackening the screw and turning the gear in the required direction.

Assembling the clock

When the movement has been completely serviced and repaired, it should be lubricated in accordance with Chapter 11 and put back into the clock case.

If, after assembling the whole clock, there is a failure of strike or chime, tackle the discovery of the fault in a systematic manner. First ensure that the gongs or bells are not holding back the hammers, then open the front of the clock, try the clock, remove the hands, try again, remove the dial, try again – and only after you are certain that no other part of the clock is interfering, take the movement out of its case. After removing the movement, examine the case for splinters of wood or anything that might have made contact with the movement, and then set the movement up on the bench and test it.

The chime side of a clock is the hardest working train, so make sure that the hammers are not being lifted when the train stops. You should have found such a fault before insertion into the case, but it is often possible to adjust when fully assembled. Simply slacken the grubscrew on the collar of the topmost gear that drives the pin barrel, move it backwards or forwards to either complete the last note of the recent phrase or pull back off the first note of the next one.

Other than this there are no special points to be watched. All the remarks that have previously been made about the location of levers, looseness of pivoting points, accurate control of the plane that the levers move through and polishing of contact areas, apply to chime-work levers and posts. So do my comments on fitting taper pins to all posts and cutting them to minimum length.

11
Lubrication and Friction

Throughout the history of machines lubrication has been a problem, particularly in relation to clocks. The usual method of reducing friction has been to use a slippery fluid such as oil or grease. Other lubricants – though not generally of much use in clocks – are water, fat and air.

Oils are the most frequently used lubricants for non-electric clocks, but those oils that were available up to the last quarter of the nineteenth century had a marked tendency to change their physical characteristics with age and temperature. By comparison with the life of a well-built clock, these changes occurred rather rapidly, so that much of a clock repairer's work lay in 'oiling' his customers' clocks on a regular basis so that fresh oil could be introduced to the bearings thinning the original oil, and then wiping away dirty, sticky and dust-laden oil with a paper spill. The regularity of this service varied between two and five years, unless the repairer was fortunate enough to have the custom of a great house on a yearly basis. Five years was really too long, and often led to the customer adding extra weight to his weight-driven clocks as the oil gradually stuck the pivots in their bearings.

Viscosity is the vital characteristic for clock-oils since bearing pressures are low and it is the consideration of 'thickness' and the ease with which a pivot will turn, that claims most attention. Oils intended for platform escapements cope with areas of design that are much more critical. Despite the low levels of energy used to drive these small mechanisms, pressure at the bearing can be quite high, pressure being load divided by the area supporting the load. The size of platform pivots is very small and the consequent clearance or shake in the bearing is small too, leaving very little room for lubricant.

A clock-oil is therefore chosen for its ability to lubricate, for a low tendency to evaporate, for low viscosity and a low tendency to react with the air and with the material of the pivot and bearing. Most oils gave sufficient lubrication for clock bearings, but until the advent of mineral oils, in the last quarter of the nineteenth century, the choice lay only between animal or vegetable oils. In the 1877 edition of *Encyclopedia Britannica,* Lord Grimthorpe, designer of the Westminster clock – Big Ben – states that olive oil was most commonly used, but that animal oil was best and 'sperm oil finest of all'. He also gives instructions for separating off fine oil from neat's-foot oil – presumably deciding that sperm oil was too expensive for common clock repairers. Equal parts of oil and water were well shaken together and then left to stand in a glass vessel for several days, the fine oil was then drawn off the top. He does not state how long this lubricant's viscosity would remain constant when exposed to working conditions. Incidentally, neat's-foot refers to oil extracted from cattle, 'neat' being derived from the Scandinavian for oxen. Besides these recommended oils, there were those that came most freely to the enthusiastic 'oiler's' hand – poppy oil, sunflower oil, fish oil, goose grease; though I doubt whether anyone, nowadays, would have recourse to any of these.

Mineral oils, on the other hand, can be so constructed as to remain fluid for a very long time and not to change greatly as a result of reaction with either metal or oxygen. There are a large number available from normal materials suppliers for clock repairing and specified according to the type of movement they are intended for – turret-clock, long-case, or carriage-clock trains, and platform escapements.

Apart from its ability to ease the friction of a clock for a long period of time, a good oil must avoid etching the metal of the bearing to an extent that produces a visible stain, usually green. The products of such a reaction are solids, the surface produced is rough, and the whole effect is at least as bad as having the oil thicken due to evaporation. A recent development in clock lubrication has been the availability of synthetic oils. They are even more resistant to drying out than modern mineral oils. The oils I use are:

Platforms and fine pivots	Synth. 60 cs
French movements generally	Synth. 80 cs
Small pivots in long case	Synth. 90 cs
Spring barrels	Synthetic grease with PTFE

(also available in a penetrating form)
"cs" = viscosity measurement in centistokes.

Mineral oils are still as good as ever of course, but synthetics give that extra certainty when lubricating clocks.

Other, non-fluid lubricants can be added to oils and greases, in particular graphite, molybdenum disulphide and micronised PTFE. All three are good and can be obtained already dispersed in a medium of oil or grease. The only disadvantage of the first two is that they are black or at least dark grey, and apart from staining the hands, they give the impression that wear is taking place in the spring or bearing. PTFE on the other hand is non-staining and very easy to apply without making a mess. When carried in a synthetic grease it is unaffected by any temperature that a clock movement might be expected to experience. Possibly the worst thing that an oil can do is develop a skin and, eventually, a varnish-like quality. This results from using raw vegetable oil, and many old clocks show evidence of such lubrication from kitchen sources.

If the train is well designed and all the moving parts including the pivot holes are well polished, a longcase movement will 'go' without lubrication except on the wiped surfaces of the pallets in an anchor escapement. The recoil of this escapement results in very high friction at the pallet face, and is probably the only place in any clock where pressure can result in metal-to-metal contact by breaking through the lubrication film. Usually a dry movement can only be achieved when relatively high counts of teeth on wheels and pinions are employed; most longcase and bracket clocks use eight-leaved pinions and the inefficiency of this train results in weights of more than 3.6kg (8lb) (or the spring equivalent) being fitted. Friction is a function of load. In other words if the overall load imposed on the train of a clock is raised because of the need to overcome train losses, the friction at the pivots – and every other rubbing surface – is also increased.

Friction, expressed as pounds or grams, is calculated from the simple expression:

Friction = Weight × Co-efficient of Friction

If the force needed to 'shear' through a film of lubrication that is clinging to pivot and bearing is ignored, friction is independent of the area of the part involved. The coefficients are numbers only, with no unit of weight attached and generally follow the following pattern: polished dry metal on polished dry metal 0.2; the same parts with a light oil 0.15. However, if the load on the bearing surface is very light, the coefficient of friction for a dry bearing may well be less than that for a lubricated one. We are discussing friction as a load; obviously if two diameters, one larger than the other, carry the same frictional load at their outer surfaces, the larger diameter will require more torque to turn it. It is a question of leverage. Since the laws change with transition from light to heavy loads, it is quite possible to have a set of circumstances on a clock's slow-moving parts (they *all* move slowly) that results in a well-made clock with polished pivots and a high count, having a higher frictional loss after lubrication than before any oil touched it.

Wear in a movement is almost always caused by grit, bad meshing of wheel and pinion, or too heavy a weight. Excessive oil on pivots will result in a 'collar' of oil that cannot help but collect dust and produce an abrasive slurry of oil and grit. Clearance over the top of the pivot and within the pivot hole ensures that progressively larger grits slide into the cavity and under the pivot.

Insufficiently polished pivots have a slight pumping effect, the helical marks on the filed surface acting as a fine screw so that the oil moves slowly through the hole carrying dirt and grit with it. Any slight sideways movement of the hand whilst filing will produce partial helices. The pivot will often be seen to be dry at one end and oily at the other, if fresh oil is introduced at the dry end – whether shoulder or oil-cup – without cleaning the entrance first, fresh grit will be carried inside. Limiting the supply of oil in the first place will prevent a flow through the bearing. Strangely, this point about over-oiling resulting in increased wear rather than simply being unsightly, does not often get made.

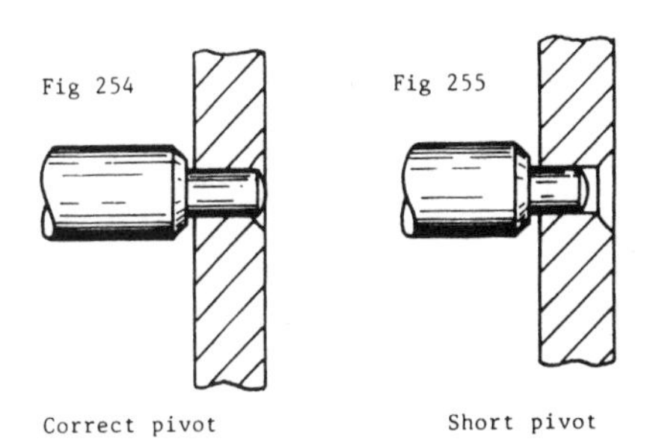

A pivot that protrudes out of the oil-cup will catch grit when oiled, rather like a flypaper (Fig 254). On the other hand if the pivot is too short and lies within the thickness of the plate, grit will enter the pivot hole directly (Fig 255). The distance over the shoulders of an arbor is shorter than the distance between the plates by a small amount called 'shake'; if this amount is added to the thickness of the clock plate it makes a very good guide to a reasonable pivot length (Fig 256). Shake is very much a matter of personal foible, for it must accommodate possible flexing of plates and the tolerances of whatever instrument the maker is using to measure over the shoulders; but if the above rule is followed, pivots will never be too long or too short so long as the shake is smaller than the depth of the oil-cup (Fig 256).

Jewelled pivots are normally found in the platform escapement of clocks and not in the train (Fig 257). There are exceptions but, outside the field of chronometers, they are very few. Basically, a jewelled bearing consists of a short cylinder of hard glass, or a precious stone (jewel) of low quality. The cylinder usually has a waisted bore and an oil-cup at one end; it is mounted in brass. Vertically mounted arbors have end-stones, also mounted in brass, that sit on the outside of the bearing and present a hard flat surface to the end of the pivot. Fig 257a shows a typical jewel bearing and its mounting.

A similar arrangement using a hard steel end-plate proves very useful on train pivots. The pivot is made long enough to ensure that the arbor shoulders never touch the inside of the clock plates (Fig 257b). The ends are rounded on the lathe so that they remain concentric, and the overall distance from end to end of pivots and arbor measures less than

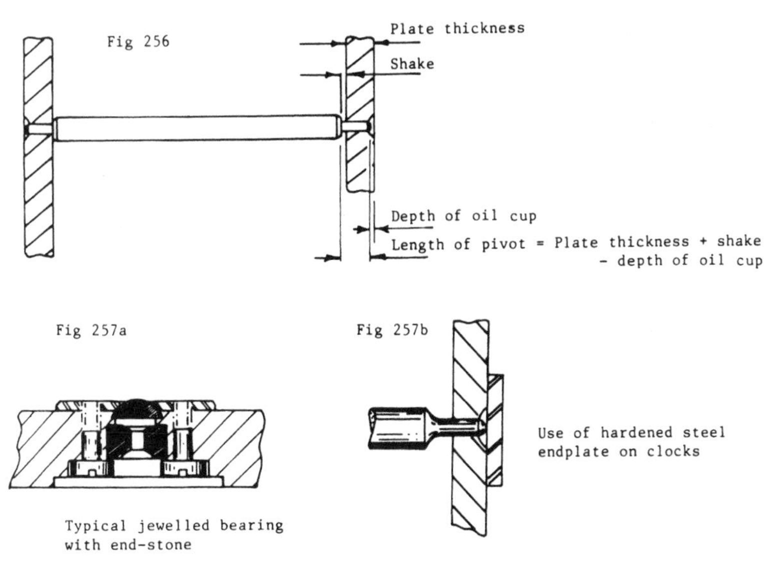

the outside width of the clock plates. The amount 'less' is the shake which, on a longcase, I make between 0.4mm to 0.6mm (0.015 to 0.025in), and on a smaller clock similar to a French mantel clock about 0.25mm (0.01in). The shake, in effect, must be less than the depth of the oil-cups, otherwise the pivots will withdraw into the pivot hole. It is no longer necessary to have square shoulders so that the stress-inducing sharp corner can be eliminated and the pivot diameter can be swept up smoothly to meet the outside diameter of the arbor. Clearance between the shoulder and inside of the plates is no longer critical since it does not control the position of the pivot end; just make sure that neither the shoulder or its alternative smooth radius touch the inside of the pivot hole.

The advantage of this end-plate is that, by moving the point of friction with the plates (axial pressure) from the annular area of the shoulder to the end of the pivot, the amount of friction is not lessened, but the radius that it operates at, is. On a pivot of 1.5mm (0.06in) diameter and a shoulder of 2.3mm (0.09in), the friction is applied at a radius of about 1mm (0.04in). With an end-plate fitted this same load is applied at a radius that cannot be calculated without being more specific about the end form, but is very considerably smaller than 0.5mm (0.02in). In experiments carried out on a longcase movement, the necessary driving weight was reduced by 60 per cent.

From this it was assumed that fully half the frictional losses in a simple clock occur at the shoulder and, of course, within a very few minutes of starting any clock, all shoulders are fully displaced to one side or the other as a result of minute but unavoidable misalignment of the plates. An additional advantage to the use of end-plates is that the oil-cups, and the oil, are protected from dirt and dust.

Arbors with extensions present a problem if end-plates are to be fitted, for it is not possible to apply end pressure to the centre arbor on the extension that carries the minute hand. Two methods can be used to lessen the losses at the centre arbor. The first is simply to put the pivot hole at the back, out of alignment, so that the arbor is held back by gravity and the winding action mentioned before. The other method is to provide a hardened steel collet only slightly larger than the pivot at the extension end; and having a polished face to contact the pivot shoulder. The reduction in friction is not so great as that achieved by end-plates, but the radius is shortened somewhat, and I believe the coefficient of friction for hard steel on hard steel to be lower than that for hard steel on brass. For one thing, the dust that falls on the surfaces does not become embedded in one half of the bearing and scour the other half.

It is important to keep the dimensions of a conventional shoulder to a minimum. A compromise between the need to reduce frictional losses and the plate's ability to resist the shoulder boring into it, has to be reached. If, for instance, the shoulder on a longcase escape arbor was only a hair's breadth high, it would very shortly drill its way into the brass as a result of the edge crumbling around the hole. Dust and metallic particles would raise the frictional losses to a point where total failure of the clock might easily occur. As a purely arbitrary measure I make my shoulders about 1½ times the pivot diameter. This measurement is taken at the chamfer and is, naturally, smaller than the arbor diameter (Chapter 4, Fig 63).

Other bearing materials are used on occasion – wood, Duralumin, ball and needle bearings and various plastics. The first, wood, recurs throughout horology, with lignum vitae, rock maple, oak and other hardwoods being used. Lignum vitae needs no lubrication, but the other hardwoods are usually soaked in oil before being fitted into the clock. Since wood exhibits greatest resistance to crushing when the pivot runs parallel to the grain, and the grain swells with lubrication and the effect of climate at right angles to the grain, it is necessary to have much greater clearance in the pivot hole. Consequently, as far as I know, only the larger movements make use of wood.

Duralumin has the facility to harden with age, due to the inclusion of a small amount of copper in the alloy. At one time, mainly from the 1950s to the 1970s, there was considerable interest in the use of the metal for clock plates and wheels but this seems to have died, possibly because of its appearance after exposure to some household atmospheres – it assumes white powdery lumps and is unpleasant to handle. Its use was mainly in industrial clocks, or control gear. Such a wide variety of light alloys has, however, been developed for industrial, maritime and aviation use, that there are undoubtedly some that would resist even a British household.

I have never cared for ball-bearings in very lightly loaded and intermittently moving conditions. The ball has to be restrained from sideways movement either by a cage, or by running in a groove and a cage. The friction that must develop as a result of various parts of the individual balls rotating at different speeds according to their distance from the centre of individual rotation, and contacting a part of the race that is either stationary or moving at yet another speed, seems to be greater than that developed by a polished cylinder of constant diameter turning in a polished hole – in other words, a conventional pivot. Needle bearings, however, only make contact with the race at their outside diameters, and with the cage by their pointed ends. Where the bulk of the bearing can be comfortably contained, they should show a much lower frictional loss than conventional pivots.

Plastics, PTFE (polytetrafluorethylene) in particular, can have very low coefficients of friction, and the problems of using them in clocks are mainly concerned with fitting them into the plates. Most of the plastics compress rather easily, and consequently there is a risk of losing dimensional stability and accurate location. The only clocks I know that have plastic bearings are 'throw-away', and more expensive to repair than to replace. Wheels and pinions are also made of plastics in this type of clock, being moulded and not cut. Apart from the fact that I do not like the look of plastic wheels, I must admit that they would probably wear as well as brass in a light movement.

Friction is affected by the wheels and pinions of a clock – the efficiency of meshing, the form of the teeth, surface finish and material used, all vary the amount of energy lost in turning the train. Clock gears rotate relatively slowly and the form of their teeth does not have as great an effect on losses as the form of a fast, heavily loaded gear tooth. It is very bad practice to 'lubricate' clock wheels. There should be no need to introduce an oil film between the teeth of wheel and pinion, for the actual pressure is very low. That at the great wheel on a longcase carrying a 4.5kg (10lb) weight – the most often over-loaded movement – must be about 44 to 49kg/cm^2 (600 to 700lb/sq in) in steady loading and double that upon recoil, and at the centre wheel 13 and 28kg/cm^2 (200 and 400lb/sq in) respectively. The result of putting oil on the teeth of clock wheels is to produce a cutting slurry, as it holds onto any dust and dirt that falls on it. I know of an elderly lady being advised to oil her clock once a week, and the repairer who came to it later found a drip-tray installed underneath the movement. Whether all this oil wore away the wheels quickly or not I do not know; it is just possible that all dust and dirt was washed into the drip-tray on oiling day!

Clock wheels are commonly made from brass, the pinions from steel, and, more often than not, the steel is worn and the brass barely touched. This apparent anomaly is created by grit being pressed partly into the brass where it is held (the brass being softer than the steel), presenting sharp cutting edges to the pinion. Most wheels in well-made clocks are hard or half-hard brass, in fact the old clockmakers often had the edges of the wheel-blanks well beaten before cutting, to harden them. However, this is entirely contrary to traditional bearing practice, which depends on a relatively soft material supporting a hard one so that any hard particles are driven *below* the surface of the soft material. A hard brass is sufficiently soft to hold the grit, but not soft enough to let it be pressed below the wearing face. The object of beating the wheel was to ensure that the tooth did not bend in use, but in the form of tooth used in medium to large clocks this never happens anyway when using an annealed wheel. Teeth bend, or are torn off when the clock suffers some trauma or other, but not in the ordinary course of telling the time. (Incidentally, wheels become annealed by accident when inexperienced repairers try to solder in new teeth.)

Many nineteenth-century cheap clocks had soft wheels and lantern pinions with hard pins. These rarely show wear on the pinion and very often not on the wheel either, yet the wheels are thinner than I like to see. All wheels I make for repairers are taken straight from half-hard sheet and cut in that condition; none have come back with bent teeth.

As already mentioned, the finish of wheels

and pinions affects the friction developed. Pinions from French carriage clocks and mantel clocks of good quality have extremely good finish, so have their pivots, and the result is that they run on light loading with very little evidence of wear. Most of the repairs, as opposed to stripping and cleaning, are the aftermath of accidents or inexperience – the commonest damage is caused by taking the movement apart with some residual load left on the spring barrel. Well-polished pinions and wheels that have had all fraze or burr removed before replacing in the train, require less energy to run than unpolished ones.

On occasion you will find that, despite having treated all the parts of the train carefully, there is a little unevenness in its running. It is permissible to let the train run fast a couple of times, by leaving off the escapement or hammer and front-work, to take out any transient lumpiness that would be difficult to find and correct. Control the speed with a finger-tip so that it does not run so fast as to be noisy. Under no circumstances should you attempt to 'run-in' a train with abrasive on the teeth. Even if cleaned well afterwards, this practice leaves scoured-out hollows in the teeth that provide pockets for dirt to pack into and increases the friction of one tooth on another, and it will prove impossible to remove embedded abrasive from the brass.

Applying the oil

Oil should be kept covered so that it does not gather dust and grit and the applicator that is used to transfer it to the oil-cups of the movement must be kept clean. A simple piece of brass wire or pivot steel with a spear-point at the end (Fig 258), will ensure that it is easy to place the oil where needed. Be sparing, the size of oil-drop that you take from the bottle will depend on the size of the applicator or 'oiler'; since small pivots take less oil than large ones use a smaller oiler for these.

Clocks that have been kept over a radiator or a fire frequently dry out; there is nothing wrong with them except the fact that the oil has ceased to lubricate. The old dried oil can be washed from the pivots by applying drops of benzine, petrol or lighter fluid on one side of the plate and soaking it up on the other side with paper. This is an acceptable way to treat a clock that, for one reason or another,

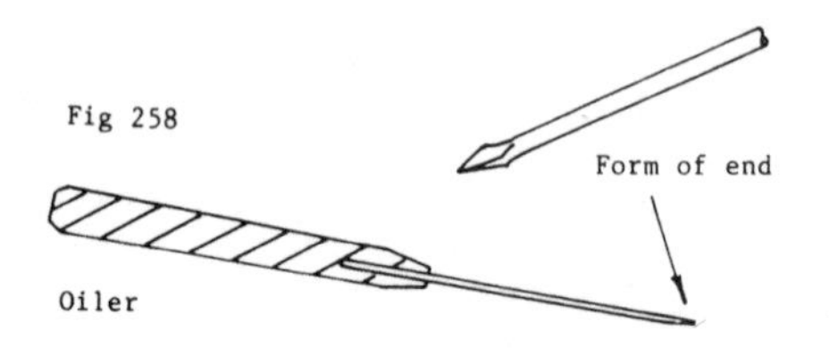

is not to be dismantled; but if it has spring barrels it will be difficult to lubricate the main springs. You will find that an old hypodermic syringe fitted with a small diameter of plastic tube will enable you to first of all wash and drain the spring, and then lubricate it. Leave the movement to drain afterwards before fitting it back into the case, for applying oil in this manner may well result in more entering the barrel than will stay on the spring. In all other circumstances the most important rule that can be applied to the lubrication of clocks is – be niggardly. Here is a check-list of lubrication points in clocks:

All oil-cups in the movement plates and cocks.
Impulse faces of escapements.
Impulse pins of platform escapements.
Escape-wheel teeth of platform escapements.
Point of contact between crutch and pendulum.
All posts for the front-work.
Mainsprings.
Fulcrums of clicks.

12
Weight and Spring Drives

The energy that is used to drive the majority of clocks (ie all mechanical clocks), is stored either by lifting a weight or by winding a spring. This chapter deals with each in turn.

WEIGHT DRIVES

The weight can be supported by gut, wire-cable or rope, each of which has advantages and disadvantages. The first, gut drive, is often used in longcase clocks. It is pleasant to work with, but if handled carelessly the catgut of which it is made will crack and develop a weak point. It is sufficiently strong to carry weights of up to about 8kg (17lb) on the pulley. A very similar line is made from waxed thread twisted into a cord. Both types are all the better for rubbing through a block of beeswax before fitting to the clock, but still beware of all sharp edges that either may contact. The brass of a winding barrel often has a sharp edge in the hole where the cord enters and is knotted. This edge will quite rapidly cut through a gut-line so that the weight crashes to the ground. I have heard of a barrel having thirty-odd knots inside – evidence of the frequency of repair.

Wire-cable is sometimes used quite unnecessarily. It is stronger than gut, but more unpleasant to handle because of its tendency to produce sharp pieces of wire that catch in the flesh. It really ought not to be used on anything lighter than about 10kg (20lb).

Chains for thirty-hour clocks often wear to the point where the pitch of the links is altered and the chain then rides over the points of the chain wheel, dropping down again with a loud 'thump' that is most unsettling. Such a chain ought to be replaced, but do not count on being able to do so without dismantling the clock. The points of the chain wheel wear, affecting the pitch required since notches worn in the points can prevent the links from dropping onto the hub of the wheel and modern chains are not made with such a wide variety of pitches as antique ones were. Some old clocks used chains with alternate links of different dimensions, so that the number of available pitches was twice the number of differently sized links – the chain wheel engages every other link. If links are being made in 3mm (0.12in) increments, the pitch increment can be 1.5mm (0.06in) by alternating links. It can sometimes be quite tedious to fit a new chain to an old clock, requiring the repairer to dismantle the chain wheel and modify its hub and points to suit the slightly different modern chain.

Rusty chains can be cleaned well by putting them into a brass or iron tube with pebbles and some sand, stoppering the ends and then rotating the whole at low speed in the lathe. Do not do this with a worn chain that still has some life in it. Clean the chain thoroughly afterwards and oil it.

Rope drives appear on very old clocks. The only thing that ever goes wrong is simple breakage or chafing of the rope. New ones have to be made up into an endless loop by a form of splice. Split each end into a number of 'tails', remove half the tails from each end by cutting out every other one, lay the tails of each end together – closing the loop – and roll to compact them until the join has the same thickness as the rest of the rope. Finish by sewing through and through with strong thread. Frankly it is the sort of job that is better demonstrated than written about.

Anchor points

One end of the gut-line on an eight-day clock is fastened to the barrel and the other – after descending into the clock case, around the pulley and up again – is fastened to the seat board. The position of the anchorage should be chosen so that the vertical lines remain untwisted when the clock is wound down. The usual cause of twisted lines is too little distance between the barrel and the seat-board anchorage; there should be space enough for the two parts of the line – descending and ascending – to remain parallel to one another. Make sure that the length of the cord is not so great that the weight can rest on the ground or the bottom of the case when the movement has fully run down; there will then be no chance of the gut coming off the barrel and becoming wrapped around the winding arbor or spindle.

The preferred method of making the seat-board anchorage is to pass the cord through a small hole in the board and then to fasten it with a clove-hitch around a small rod so that it cannot pull out of the board. Make sure that the seat board is firm and cannot rock. The case should be stable too, otherwise the clock will very probably stop when the length of the line between seat board and weight is close to the length of the pendulum rod. In this state the weights are likely to harmonise with the pendulum and begin to swing in time with it; the resultant movement of weights and pendulum bob can very often move the case of a clock.

When installing a weight-driven clock, have the lines fully unwound so that there is small chance of them becoming wrapped around the winding arbor, and keep tight hold of them until a weight can be hooked onto its pulley. The pendulum goes on last and comes off first for the same reason – maintaining the stability of movement and seat board; not all seat boards or clock cases are well made and stable.

SPRING DRIVES

Barrels

There are two types of barrel; the plain spring barrel used to drive a fusee (Fig 259), and the going barrel that carries a gear on its end called the barrel ring (Fig 260). Both types consist of a tube containing the spring and closed ends that are bored to fit an arbor (Fig 261). Most barrels have only one of these ends capable of being removed, but occasionally barrels for fusee clocks have both ends held in place with screws. The removable end is called the 'cap'.

When dismantling a spring-driven movement there is one very important precaution to be taken; run the clock down and remove either the ratchet or the 'click' that slips into the ratchet teeth and holds it wound up. Unless this is done there is always a possibility of some residual movement being left in the spring after it is wound down (one or two coils may be stuck with grit or old oil), that can be released when the clock plates are taken apart. Without a ratchet in operation this movement will do no harm, but simply turn the winding arbor; however, if the ratchet is still in place and engaging the click, all the movement will be transferred to the gear train, and very probably damage gear teeth or pivots. You will find this comment made elsewhere in this book, I make no apologies for duplication, a very great amount of damage is done to spring-wound clocks by ignoring this precaution and professional repairers are by no means guiltless. It is very easy to make an error with a spring-driven clock if it has been preceded by a number of weight-driven ones, but a habit of removing all ratchet wheels in sight will go a long way to counteract a moment's thoughtlessness. Comments on the dismantling of clock movements will be found in Chapter 2.

Some modern clocks have barrels with ends that are pierced so that the condition of the spring can be readily seen. Older examples must have the cap removed before the spring can be inspected. Almost all going barrels have a cap that is a press fit into the end of the barrel, and which can be removed either by inserting a screwdriver blade in the small cut-out that will be seen in the cap and levering it out, or by striking a piece of brass sheet with the arbor of the barrel, keeping the cap uppermost. The blow should drive the cap out of its seating and the brass sheet will avoid the arbor being bruised.

Levering the cap off will damage the rim of the barrel; you should ensure that this damage does not spread during ensuing removals by making sure that, when the cap is pressed back again, the aperture is in the same place as before, thus limiting the number of indentations in the rim. This position will either be already marked with a dot or should be directly in line with the anchor (the internal fastening for the spring), which shows on the outside.

Fig 259

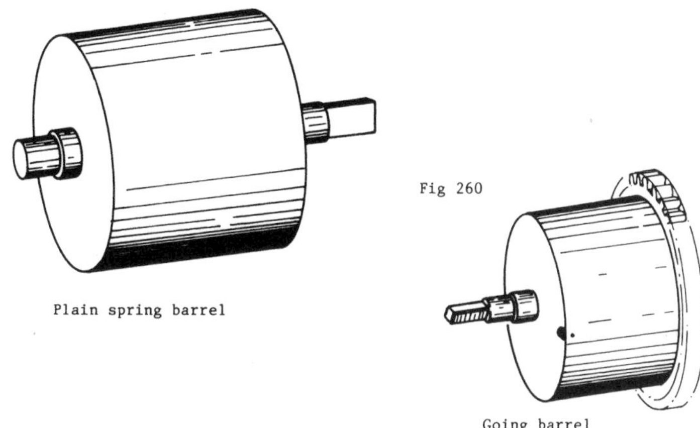

Plain spring barrel

Going barrel

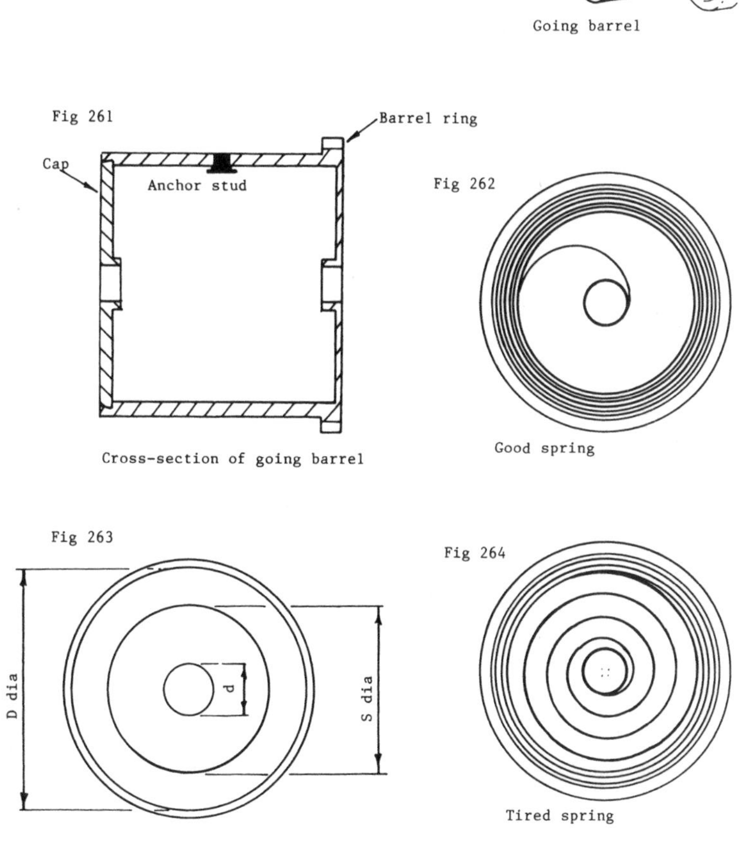

Cross-section of going barrel

Good spring

Tired spring

Barrels for fusee clocks are often provided with similar caps to going barrels. There is a cut-out in this cap, too, but it is intended to clear the knot of the gut or cable, and if the cap is levered off you will have to be very careful not to damage the knot-hole with the lever – it is better to strike the arbor on a piece of brass. Modern fusee barrels may have their caps held by means of small screws let through the circumference of the barrel tube and into the thickness of the cap.

Condition of spring

Inspect the coils of the spring, they should all, with the exception of the innermost coil, be pressed flat against each other and against the inside of the barrel tube. Only the inner coil should wind across the space between the arbor and the outer region (Fig 262). If more coils than this wind across the inside of the barrel the spring is generally termed a 'tired' spring (Fig 263). There are some exceptions, in particular many modern German clocks appear to have been designed with more space in the barrel than is needed, and the spring does not make use of the space in the most efficient way.

For a spring to make the best use of the space within a barrel it must move its coils the greatest possible distance; each coil will travel from the wound condition to press firmly against its neighbour in the unwound condition (Fig 262). The whole available area within the annulus of barrel and arbor is halved, the inner half is occupied when the spring is wound and the outer half when the spring is entirely unwound.

Since a major cause of the failure of a clock to go for its designed period – thirty hours, eight days, 30 days – is either a spring breakage or the replacement of the original by another and unsuitable spring, we ought to take a brief look at spring design.

1 The area of the barrel annulus that is occupied by the spring is:

$$\frac{\pi}{4}(D^2 - d^2) \times \frac{1}{2} \quad \text{where} \quad \begin{array}{l} D = \text{inner barrel dia} \\ d = \text{arbor dia} \end{array}$$

2 If half of the annular area is occupied by the spring, the inside diameter of the unwound spring (as it lies in the barrel) can be calculated (Fig 264). Let S = the inside diameter of the spring:

$$\frac{\pi}{4}(S^2 - d^2) = \frac{\pi}{4}(D^2 - S^2)$$

$$S^2 - d^2 = D^2 - S^2$$

$$2S^2 = D^2 + d^2$$

$$S = \sqrt{\frac{D^2 + d^2}{2}}$$

3 The barrel arbor is very often about 30 per cent of the inside barrel diameter and if we give the latter as unity (1),

$$S = \sqrt{\frac{1 + 0.09}{2}}$$

$$S = \sqrt{0.545}$$

$$S = 0.738$$

In other words, for this proportion of barrel and arbor, the inside diameter of the unwound spring is about 74 per cent of the inside barrel diameter.

4 The difference between the number of spring coils when wound and when unwound is called the 'coils of development' and represents the number of turns that the barrel can make. This number of turns should rotate the centre arbor 30 times in a thirty-hour clock, 192 times in an eight-day clock and so on. It is quicker to rotate the barrel by hand than to count the teeth and calculate.

5 If 4 does not provide sufficient turns, the spring is too thick – this dimension is called the 'strength'. If the clock fails to 'go' it is a possibility that the spring is too thin.

6 The number of coils when wound is found by dividing the radial distance from the arbor to the diameter S (see 2 and 3) by the thickness of the spring. The number of coils when unwound is found by dividing the radial distance from S to the inside diameter of the barrel by the thickness.

7 The width of the spring should be almost the same as the distance over the shoulders of the arbor – the inner shoulders, not those between the clock plates.

It should be possible, using the notes 1 to 7, to check on the suitability of any spring found in a suspect barrel, and to calculate its replacement. Spring stockists require the inside diameter of the barrel, the width of the spring, and its strength. Fit the nearest available spring that is smaller than the barrel diameter.

Removing the spring

Although the condition of the spring can be judged to a great extent whilst it is still within the barrel, it will be necessary to remove it if there is any hint of dirt, old degraded oil, or deterioration of the spring itself.

Spring winders are sold for the express purpose of removing and inserting springs, and it is claimed that a spring can *only* be taken out in this way, if damage is to be avoided. I do not agree, but it is a matter of opinion, and experience will lead you to adopt one method or the other. Winders come with explicit instructions, therefore I shall not go into their manner of use here.

REMOVAL BY HAND

You will need a napkin-sized piece of cloth, smooth-jawed pliers and a good grip. Hold the barrel in the cloth and grip the innermost coil with the pliers, shake the cloth over the pliers so that the spring, when it emerges, will be shrouded. Rotate the pliers to tighten the spring and draw the coils out *gently;* the intention is to wind the spring out, not to drag it protesting into the daylight. When the top halves of the inner four or five coils are withdrawn, it should be possible to put the pliers to one side and grip the spring with the hand, through the cloth, and twist it until it is winding out of its own accord.

The entangling effect of the cloth and your own grip will control this, so that the spring is not strained by flinging outwards suddenly. When all the coils are out, the spring can be slid off the barrel anchor – a small stud inside the barrel tube that fits into a slot at the end of the spring.

DISTORTION

If you hold the spring by the end, it should not show any pronounced 'belly'; 12mm (0.5in) on a 40mm (1.5in) barrel is acceptable but, if there is more, inspect the surface of the spring for kinking. As long as the surface is still smooth with no sign of sudden changes in curvature, the belly can be taken out by running the fingers along the spring and tilting it firmly (but again gently) to left or right, according to the direction of the distortion. Incidentally a tired spring that has clearly been in place a long time can very often be given a new lease of life, by straightening it out with the fingers. Obviously if the spring is new and become tired, there is some problem with its strength (more properly its 'temper'), in which case there is no point in straining it again in this fashion. However, all springs lose a little of their facility to recover from the strain of being wound up, at each winding, and if the observed tiredness is the result of this, the spring may very well be preserved to give many years' additional service. An antique, handmade spring is more worthwhile saving in this way than an easily replaced modern one.

The type of distortion described above – the belly – is not important, as you can very well test, for a light pressure will push the belly back again. No distortion as ineffective as that will lift the cap of the barrel, or scrape the inside. What is important is the sort of distortion that creates large forces in unwanted directions.

Consider the spring as a simple flat beam – a plain strip of steel. It is very easy to grip each end and twist slightly if it has the same length and cross-section as a typical spring. This is analogous to the bellying; very small forces are involved. If you now take a hammer and strike one edge of the strip so that it is distorted and stretched slightly, the steel strip will curve off to left or right, and it requires a great deal of effort to bring it back to its original straight condition. The distortion in a spring that has a great effect on the proper performance of the barrel is caused by damage to the edge, or kinking, or something lying between two adjacent coils forcing a tilt between them. The last can be caused by grit or fraze thrown up on the edge of the spring by forcing it into the barrel clumsily. Extracting the spring in the manner described above should not cause any damage of this sort.

Anchorages

The normal means of attachment of a spring to a barrel and arbor is by way of projections on the arbor and barrel and holes in the end of the spring (Figs 265, 266). Since the spring should not be forced to one side or the other within the barrel, it is important for the ends of the spring to be held so that no bias is introduced by the anchorage. If the holes in the ends of the spring are round or oval, the spring will slide over until these holes 'centre' on the anchorage of barrel or arbor. If the anchorage happens to be slightly out of centre,

Fig 265 Anchor on arbor

Fig 266
Spring end and anchor stud

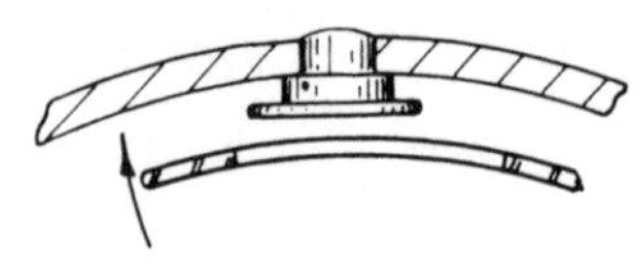

Fig 267 Form of spring end

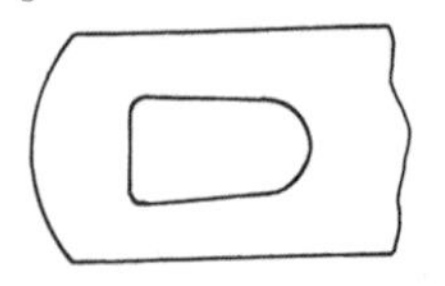

Fig 268

Spring end supported by a 'joggle'

the spring will be held out of centre too, and if there is a discrepancy between the two anchorages, twisting and possible distortion will result. As shown in Fig 267, the holes in the spring are flat-ended so that they will slide over the projection of the arbor or barrel with no bias to either side. Please note that sharp corners are to be avoided; they concentrate stress and eventually lead to cracking.

The entire surface of the spring must be smooth from one end to the other. If the anchoring holes have 'lipped' (Fig 25), they are well on the way to failure because the metal is not ductile and the lipping will be accompanied by cracks that may well be indiscernible, but are still there, nevertheless.

Pay some attention to the outer end of the spring; if it is flat or has sharp edges there is a risk of producing small particles of brass by scraping on the inside of the barrel. It should have a rounded end with all sharp edges removed. Many mass-produced barrels have an anchorage to the tube formed as a pressing projecting inwards. Since this intrudes on the spring space rather more than the normal stud, the end of the spring should be bent as shown on Fig 268 to ensure that the anchor does not dig into the next coil of the spring.

BROKEN SPRINGS

When a spring breaks for the first time, and if it is only the very end that has given way, it is quite possible to repair it by putting another hole to replace the broken end. If much more than the very end of the spring has broken off, it would be better to replace the spring, because of the shortening that would otherwise result. First cut away the broken end and then hold the end of the spring with an old pair of pliers – the ensuing heat-treatment would damage new ones by softening the jaws. Heat the end of the spring for a distance equal to half the circumference of the inside of the barrel, to a dull red. Maintain this temperature for a count of twenty or thirty, and then gradually withdraw the steel from the flame, so that over a similar count the metal loses its colour and gradually cools. Be careful not to heat any other part of the spring, particularly if the inner anchor hole is being replaced. The coils can be protected from unwanted tempering by inserting a thin strip of brass so that it shields the coils and radiates heat.

When the steel has cooled down to the point that it can just be touched with the fingers, dip it into oil – any sort of lubricating oil – and then wipe it clean. The steel is now soft and can be drilled and then filed to the shape shown in Fig 267. Remove all sharp edges afterwards and wipe down with a clean lubricating oil such as Three-in-one.

Inserting a spring

The spring winder is more useful for putting springs back into the barrel than for extracting them. However, it is quite possible to put most springs into their barrels by hand; much will depend upon the strength of the spring – and your hands.

INSERTION BY HAND

Before inserting the spring by any method, make sure that it is clean, and dry enough not to be too slippy to hold. Curl the inner coil to clasp the arbor for at least three-quarters of its circumference, and bend the end of the outside coil so that from the end of the hole onwards it curves to a smaller radius than the inside of the barrel. This will ensure that the stud or anchor enters the hole, and grips it firmly.

Make certain that you know which way the spring is supposed to be set in the barrel. It is exasperating to spend five minutes struggling to feed a strong spring into position only to discover that the arbor winds in the opposite direction to the spring. Use a cloth to hold the barrel, lay the end of the spring into the barrel and over the anchor. In all probability it will not catch on the anchor at this stage, but it is easier to lay the coils into the barrel if the stud is shrouded by the first coil. It is quite possible to get several coils in without realising that the first coil's edge is caught in the stud, and consequently it is impossible to press the whole spring down into its proper place.

The coils must go in naturally as you bend the spring into a curve, there must be no grating of the edges of one coil on another. Although you will need to supply a certain amount of pressure, the main effort should be in bending the spring to the required curvature. Take especial care with the first two or three coils; until they are seated within the barrel there is a great tendency for the spring to fly out again at the slightest relaxation of your hand. After this point, you will find that you can take a rest and hold the spring in the barrel with one hand while the other stretches itself free of cramp.

The last coils can be pressed home and then lubrication applied in the form of four or five drops of oil or a liberal amount of synthetic grease, preferably a penetrating grease in a carrier that later evaporates and leaves the lubricant evenly dispersed throughout the spring coils. If the anchor stud and the hole at the end of the spring are well matched, the spring will engage the barrel satisfactorily as soon as it is wound up. If it is convenient (or even possible), wind the spring before assembling the movement to check that the spring does engage.

INSERTION ON THE LATHE

Springs can be very strong indeed and quite beyond the capability of the normal person to insert by hand. Figs 269 and 270 illustrate a method of winding the spring if you possess a lathe with a geared headstock. I specify a geared headstock because this will offer resistance to the spring, and prevent it unwinding immediately you take your hands off it. In the absence of such a lathe a simple ratcheting system or belt brake (Fig 271), could be used.

You will need to make the supports for the winding arbor as shown and a hook to suit the hole in the end of the spring, fitted with a T-shaped handle. Grip the non-bearing part of the arbor in the three-jawed chuck, slip the spring into place and anchor it, slide the barrel over the support ready to accept the coiled spring, and locate the arbor in the hollow end of the support.

Now the lathe can be used to wind the spring – with power if you are confident and have someone with you in case of accident; by hand if you are on your own or cautious. Hold the spring by the T-shaped hook and wind the spring until the coils are small enough for the barrel to slip over. At this point you must make sure that the spring's outer coil goes past the anchor in the barrel; you can do this by tilting the barrel slightly. When the first coil is safely past the anchor, release the backgear or the brake on the headstock so that the chuck is wound back by the spring until the coils open up to fill the barrel. Alternatively, the spring can be held at a diameter that will enter the barrel by wrapping a piece of wire around it and fastening it.

Remove the barrel, spring and arbor from the lathe, unhitch the T-hook. (A wired spring can be dropped into the barrel now and the wire slid off as the spring is pressed into it.) Use a piece of brass or copper to press the spring into the barrel, whether it is wired or simply as taken from the lathe. These metals are soft enough not to damage the spring, but will not scatter dust and chippings as timber would.

REPLACING THE CAP

The cap can be put in place as soon as the spring is down to the level of the open end of the barrel. Small barrels can usually have their caps pressed on by gentle taps from the handle of a hammer or similar implement, larger ones may need to be placed in the jaws of a vice (suitably guarded against damaging the brass), and squeezed into the barrel gently. Remember that the small hole in the edge of the cap that was used to remove it, must be put back in line with the dot – if there is one – or the anchor stud.

Caps should fit firmly and not rattle. If the cap is loose, lay it on a smooth block of brass and tap around the outside with a smooth-faced hammer so that its diameter increases.

Arbor bearings

The hole in the barrel ring that accepts the winding arbor must be a good fit on the arbor, similar to the type of fit that is needed on any other pivot. If it is oversize, the mesh between the barrel ring and the next pinion will be affected and may well result in a clock failure. It is sometimes thought that the hole in the cap is not important since it does not greatly affect the mesh of the train, however it does allow the barrel to tip over in relation

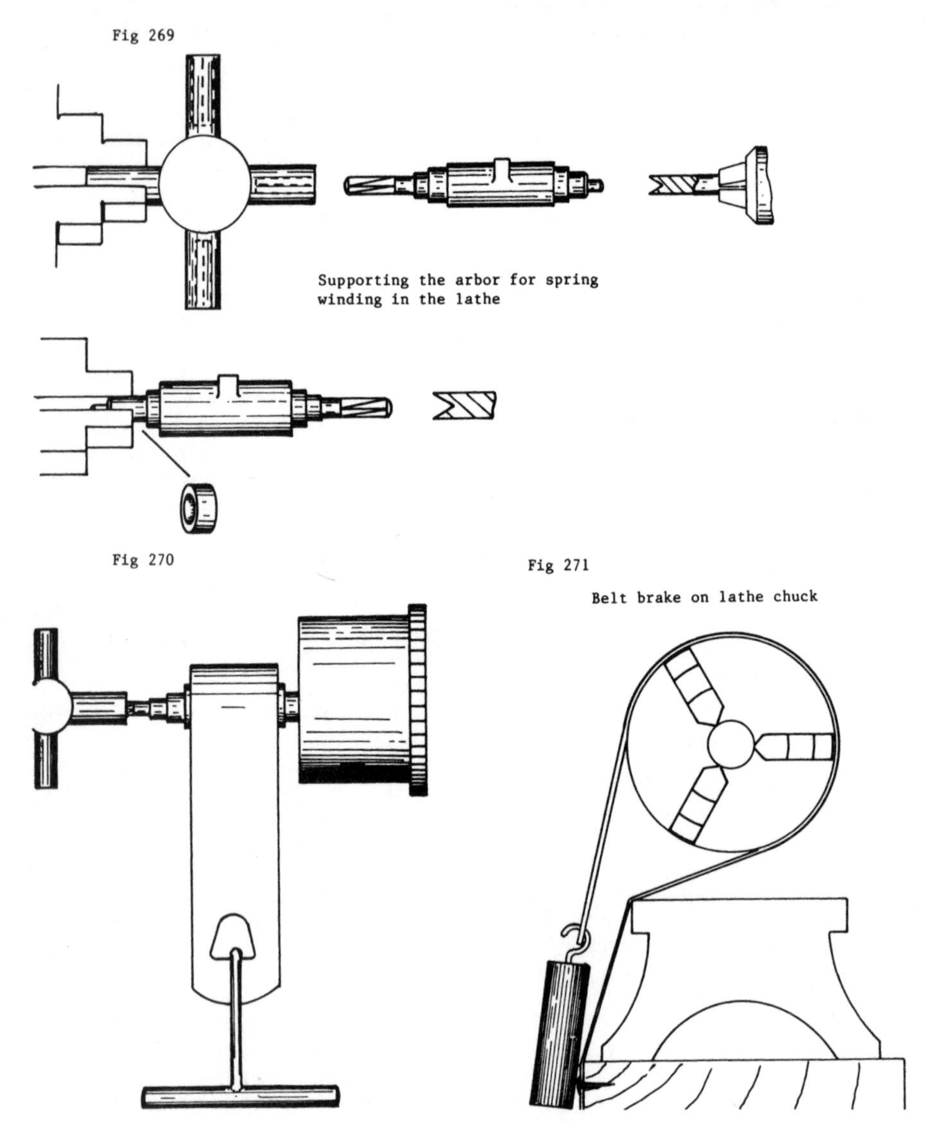

Fig 269

Supporting the arbor for spring winding in the lathe

Fig 270

Fig 271

Belt brake on lathe chuck

to the arbor and this distorts the spring. Distortion of this type will cause the spring to scrape the inside of the barrel's ends, and may force the cap off; it will certainly produce small chips of brass that will cause faults in the future. Both holes can be bushed in the same manner as the clock plates.

Open springs

Open springs are easy to inspect and clean in situ. There are two types – riveted or closed loop, and open loop, the loop being the anchorage for the outer coil onto one of the pillars of the movement. To remove either type, prepare a strong steel clip from stiff wire of 8 or 10 gauge, ie about 3mm (0.125in) diameter; it should form slightly more than a half circle of the diameter that would contain the spring and allow it to be removed without interfering with the rest of the movement. Wind up the spring until the clip can be slipped over it, and then let the spring unwind into the clip until it will unwind no further, and has no effect on the train. The movement can now be dismantled. However, if the spring is broken you will not be able to use this method, and must try to draw the spring out of the movement sideways until there is no chance of the arbors being snagged when dismantling takes place. Short rods, pivot-steel for instance, can be passed through the cut-outs of the plates to isolate the spring from the rest of the movement. If you cannot do this, just grit your teeth, undo the two nuts or taper pins furthest from the spring and loosen the others, so that the smallest pivots may be unshipped without letting the spring free to batter the arbors. Simply be as careful as possible and use any means that seems reasonable to restrain the spring when the plates come apart.

A closed-loop spring ought to be wound into a clip and then assembled with the other parts. An open loop can be fed into the assembled movement from the outside, winding the arbor until the loop can be slid over the pillar. Sometimes it is not possible to use a clip on the riveted loop and the following method can be adopted. It looks untidy, even a little cruel; but it works, and does the spring or movement no harm.

Hold the spring with both hands close together, move them apart and along the spring until you are holding both ends in their proper relationship to one another; the mid-portion will curl over itself and look terrible, but as long as you proceed slowly you will do no damage. Hook the inner coil onto the arbor, and hook the outer coil onto the pillar, release the spring slowly and assemble the movement. Have someone hold the movement while you wind the spring slowly onto the arbor, taking time off every so often to ease out the twisted spring; and carry on until all is wound in.

FUSEES

The fusee is seen more often in British clocks than in any others, its purpose being to provide the train with a constant torque (load x lever arm) so as to achieve better timekeeping. It is expensive to make, it increases the cost of maintenance and, as regards producing a good timekeeper, has rather less effect on accuracy than a good escapement, particularly if a well-made modern spring is used (Fig 272 shows a typical fusee and spring barrel). Nevertheless, it is often found in the best quality clocks, and is a much-prized feature. Over the centuries the curve of the fusee has altered, early examples have a rather flatter profile than more modern ones. This reflects a better understanding of the mathematics involved, the most recent contributions having been made in the last thirty years by such notable horologists as A. L. Rawlings.

There is little for the repairer to worry about concerning this part of the mechanism, and this is covered by paying attention to the pivots, the internal ratchet, and the cord or chain. The first of these is dealt with in Chapter 4 since these pivots are no different to any others.

Internal ratchet

The fusee must be taken to pieces before the ratchet and the associated click and spring can be inspected. Two methods of housing the ratchet are employed. One makes use of a ratchet wheel sunk into the body of the fusee and pinned there, and a click or pawl that is mounted on a simple great wheel. The other method reverses this; the fusee is plain ended, carrying a click on the flat face whilst the great wheel is bored out to accept the ratchet wheel – this too is pinned in position. It does not matter which method is found when the fusee is dismantled, the major points are the same.

To begin with, the fusee and great wheel are held together by a keyhole washer that is fitted into a groove on the arbor; this arbor is a drive fit in the fusee and is *not* removed (Fig 273). Remove the pin or screw that prevents the washer sliding sideways across the face of the great wheel, and thus disengage it from the arbor. The wheel can now be lifted off the fusee. Make sure that the teeth of the ratchet wheel and the end of the click are sharp, and fit each other. The click spring must be capable of holding the click against the ratchet teeth. Clean out all old grease or oil, and note that the ratchet wheel lifts off its pins and will only go back onto them in one way without binding on the arbor. The proper position can be found by marking the wheel before removal or noting any existing marks. Grease may be used in this assembly, since lubrication can only be carried out when the type of lubricant last used can be clearly identified or removed entirely. The major advantage of grease for this item is its ability to hold the parts in correct location during assembly.

Cords or chains

Cords may be of wire or cat-gut, and the fusees that they are fitted to usually have relatively shallow grooves that are round-bottomed. The choice of wire or gut should be made on the basis of age – the older clocks used gut or chain – and the tension induced in the cord by the spring barrel. The tension may be checked by assembling the spring barrel in the clock and discovering what weight it can lift when fully wound. Cat-gut can withstand the effects of lifting about 4kg (8.8lb) direct load if it is of the size used in most British clocks, ie 1.5mm (0.06in), and about 2kg (4.5lb) for 1mm (0.04in) diameter. Loads greater than this will require the fitting of a steel wire.

Chains are found only on fusees with flat-bottomed grooves which are somewhat deeper than cord-driven examples. Quite clearly the chain must be flexible, therefore soak it in paraffin (kerosene) and oil, and then run it backwards and forwards over a polished steel bar. Do not use wood or brass or you will produce dust or chippings, and do not bear down heavily; all that is required is to work the links until they are free. Then clean and oil the chain.

Ensure that the groove that the cord or chain lies in has not been damaged, and that the sides do not grip it in any way. The fusee should allow the cord or chain to unroll smoothly, with no tendency for it to stay in the groove beyond the point that is touched by a tangent common to the spring barrel and fusee. Correction may be made by using a small flat file on the damaged groove; you may find that the groove walls have been pressed sideways over the years. Rotate the fusee by hand to do this work; if it is held in the lathe chuck do *not*, under any circumstances, be tempted to use power to rotate the work. Clean thoroughly after polishing with crocus paper.

MAINTAINING POWER

When a fusee or a weight-driven clock is wound up, the going train comes to a halt unless maintaining power is supplied to keep it in motion. The thirty-hour clock with an endless chain passing over two chain wheels (Huygens' loop) already has maintaining power in its method of drive. Other clocks employ a spring device to turn the great wheel when the pressure is taken off the drive.

Vienna regulators, British dial, bracket clocks and longcase regulators, are the main users of a system of maintaining power invented by Harrison in the eighteenth century. This system has a ratchet wheel interposed between the great wheel and the fusee (Fig 274), and a long pawl supported on an arbor, rests on the teeth of the ratchet wheel. The drive from the fusee is transmitted to this ratchet wheel by means of the normal click, which in this instance is attached to the face of the ratchet; the ratchet slips under the long pawl and passes the drive to the great wheel by means of one or more springs at the interface. When the fusee ceases to drive as a result of being wound up, the long pawl holds the ratchet wheel in position, and the springs release their stored energy by continuing to drive the great wheel for a short time. The springs are simple in design, and replacements may be made from spring wire or strip according to the original design, quite easily.

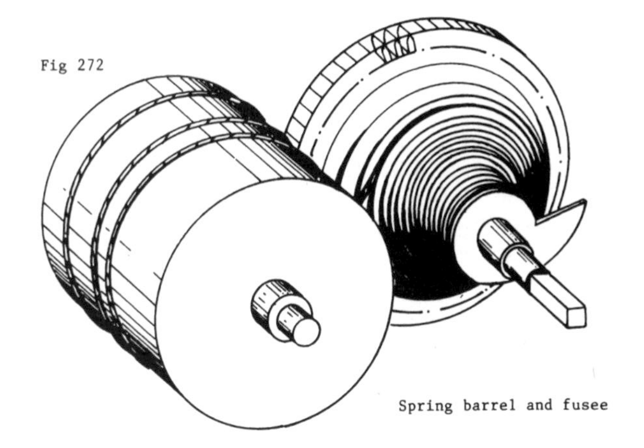

Fig 272

Spring barrel and fusee

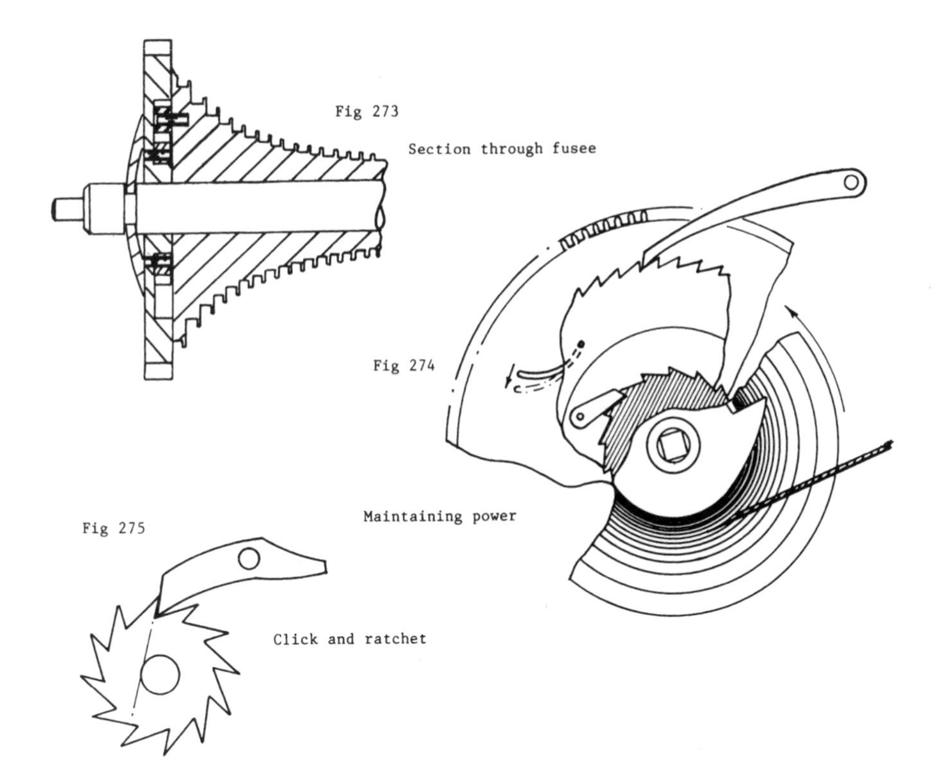

Fig 273

Section through fusee

Fig 274

Maintaining power

Fig 275

Click and ratchet

RATCHETS AND CLICKS

Ratchets are used in clocks in fusees and maintaining power, as we have just seen, and also on the winding arbor of spring-driven clocks. Since this small mechanism is all that stands between the winder of a clock and damaged fingers and/or clock, it will do no harm to look at it with a little more care and attention.

The working face of the click should fit neatly against the opposing flank of the ratchet tooth, and any radius above about 0.15mm (0.006in) that has been worn onto it should be removed. If the teeth of a ratchet are badly damaged it will probably be cheaper to replace the whole wheel; but if the old piece is to be retained, the teeth can be cleaned up with a barette file which will cut a sharper corner than the standard three-square or triangular one. Keep the working face radial to the centre of the wheel or slightly raked (Fig 275), so that the click is drawn into the tooth rather than being pushed out. Polish with fine emery cloth on a metal backing strip. The click, too, must be polished after hardening and tempering to a blue temper.

When the damage is so great that teeth are broken down or even broken off completely, a rectangular seating should be cut for a new piece of brass to be inserted, soldered and then cut to correspond with the form of the other teeth. Half-hard brass is best, but soft brass that has been raised to red heat and cooled can be made half-hard by hammering it until its thickness is reduced by about 10 per cent; any stress will be relieved by the act of soft-soldering.

Click springs

Always copy the old click spring rather than simply buy a replacement. Many replacements from suppliers will exactly match the broken spring, but when this cannot be done new ones can be made from brass or steel, according to the original metal used.

Curved brass springs such as appear on British clocks are best made by cutting the brass to the straightened form, and then curving it by beating along one edge, thus hardening it at the same time as you match it to the curvature of the wheel that it is attached to. Spring-steel clicks can be filed from silver steel. It does not take long, and the result is much better than an attempt to bend old spring steel to match the form of the L-shaped foot that the screw passes through, to hold the spring in place.

13
Tools and Equipment

It is difficult to make a comprehensive list of all the tools that a clock repairer or restorer would use, since repairers often make tools to suit themselves, and not necessarily to a recognised design. Even a 'basic' list is difficult because it depends on what work the repairer decides to limit himself or herself to – simple adjustments and re-surfacing of pivots, replacement of hand-made parts, gear cutting or whatever. In the past, clocks have been made with very simple hand tools such as files, drills, saws and broaches. Clearly it is possible to repair clocks within the same limitations, but it would demand a great deal of the craftsman's time, and the development of considerable skill. The following is designed to show the approximate order in which tools can be added to your own workshop, and give a logical progression to the type of work that can be carried out there.

The asterisk at the beginning of some of the sections is an indication of a basic or 'key' tool which is essential at that level of clock repairing.

***Pliers**

There are two types of joint for pliers: box joint, and lap joint. The former is the most expensive and gives the greater strength, however, I have never had a pair of pliers fail because of the joint shearing or distorting and, as a result, I buy the cheaper lap joint. The two types are illustrated in Figs 276 and 277.

Pliers are required for removing taper pins from clock pillars and posts, for holding or manipulating clock parts when assembling, and for bending or cutting materials. For the first of these tasks a flat-ended pair is needed with a groove cut across the end of one jaw; this is a job that you usually have to do for yourself with the corner of an emery stone. An overall length of 100mm to 150mm (4in to 6in) is suitable and the jaws ought not to be slender or 'snipe-nosed'; they would not be strong enough to remove partly corroded pins from elderly clocks (Fig 278).

The pliers that are needed to hold arbors and other parts, when assembling a clock movement, need slender jaws. Round-nosed pliers *can* be used but I prefer long, flat-nosed ones because the width of the jaw gives greater stability to the part held, and the actual thickness is less than the round-nosed type, allowing greater ease of manipulation within the confines of the movement. No great strength is demanded of these pliers, and it does no harm to temper them to a dark blue, and stone all sharp edges to reduce the chances of marking the clock parts. If the only pliers that you can buy for this job have serrations on the jaws, grind them smooth first.

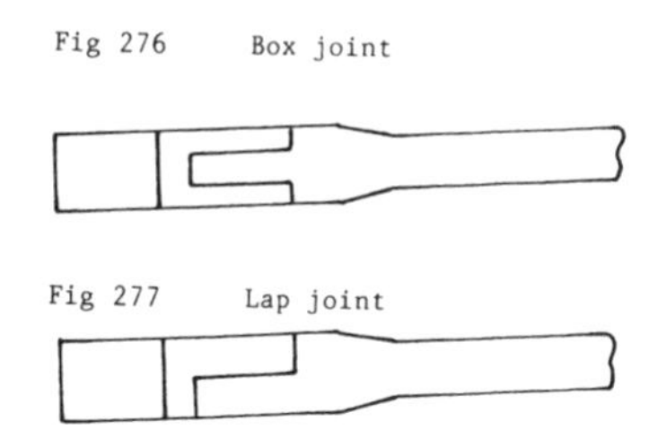
Fig 276 Box joint

Fig 277 Lap joint

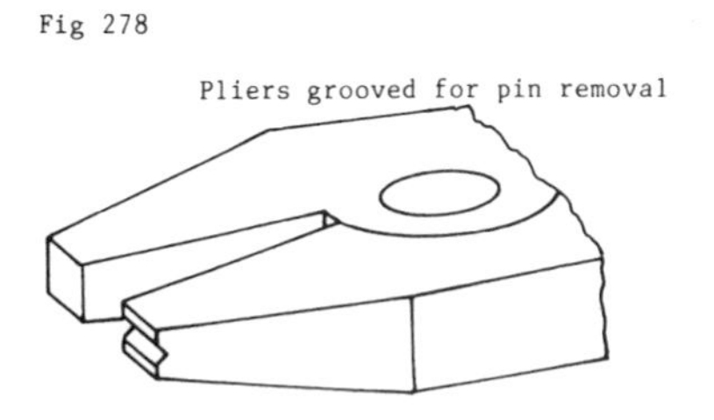

Fig 278

For the mechanical task of bending wire and for adjusting the 'set' of levers, simple robust pliers 150mm to 200mm (6in to 8in) long will serve very well – broad jaws that will not distort when strain is put upon them, typical of the pliers to be found in an electrician's or a mechanic's toolbox. Two pairs of these are advisable if you are intending to use one pair for holding work during heat treatment or soldering. These jobs will very quickly ruin a pair of pliers for any other work; the jaws become soft, and the joint and general area of the jaws will be corroded by the effects of heat and fluxes.

Cutting with pliers requires a good pair of side-cutters, and possibly a pair of end-cutters. In the main these will both be used for cutting taper pins that are too long, and pivot steel for springs and other small parts. Most of the cutters that one sees in clock-tool catalogues seem to be limited to about 1mm (0.04in) brass wire so far as capacity goes. My own cutters are electrician's capable of severing 2.5mm (0.1in) of copper or brass. They open wide enough for me to use an emery stone on the edges to sharpen them from time to time, and remove the notches worn in by pivot steel, and the side has been ground back so that they will cut flush (Fig 279). End-cutters are for cutting at right angles to the side-cutters; one can manage without them.

Fig 279a Sharpening side-cutters

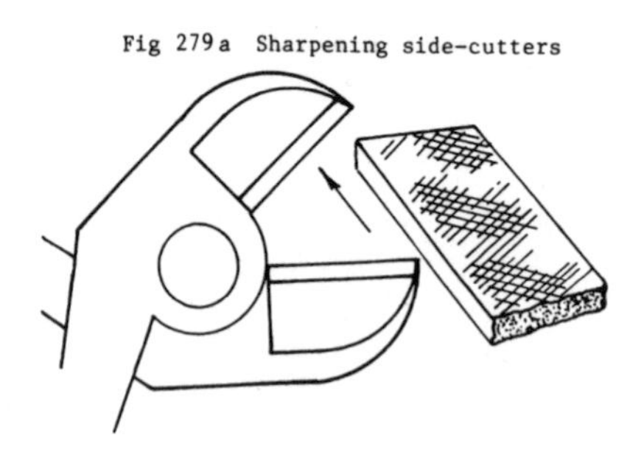

Fig 279b Sharpening side-cutters

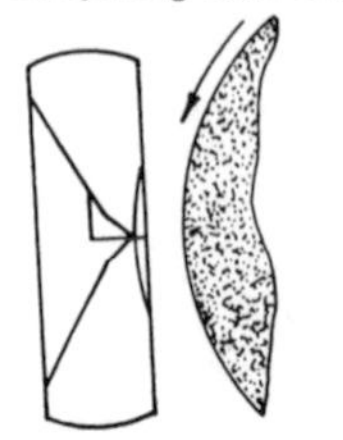

*Screwdrivers

Collect a good range of screwdrivers, varying in length from 50mm to 200mm (2in to 8in) and in breadth of blade from 1.5mm to 9mm (0.06in to 0.4in) The most common pattern is the so-called 'London pattern' – a somewhat archaic term – which has the blade in-line with the handle. In the case of the larger screwdrivers the handle will be the familiar wooden or plastic one; smaller screwdrivers ought to be clockmaker's type with a cylindrical handle, and a hollow at the top for a steadying forefinger to rest in. Since space is often confined, it is worthwhile making one or two examples of the older pattern – L-shaped or T-shaped – that can be used in awkward situations (Figs 280, 281). Use silver steel and harden and temper to dark amber.

The end of the blade should be hollow-ground for use on screws made since the middle of the nineteenth century since most screw slots from that time on were cut by saw, usually in very large quantity. Prior to that time the slot was produced with a file, and is not parallel-sided; keep a screwdriver with a tapered blade if you are likely to come across many antiques of this period (Fig 282). A useful design for a multi-bladed 'driver' is shown in Fig 283.

*Tweezers

You will find innumerable occasions when parts can only be picked up or moved into place with tweezers. They are substitute fingers, smaller, harder and with clearly defined ends that do not obscure the manipulation of small parts. Choose them as they become needed. Many household types are very useful, and the fact that they often become magnetised is an advantage. However, you will definitely need a non-magnetic pair or two, which can be bought from clock-tool suppliers. As an emergency measure, two 'fingers' of a set of feeler gauges can be taped together at one end and the other end shaped to suit the job in hand. The thinner fingers, 0.075mm to 0.15mm (0.003in to 0.006in), are most useful. I have a set in use that was made a year or two ago for a delicate job, and although they do not look pretty they have proved invaluable.

*Brass brush

At first sight this may not seem particularly

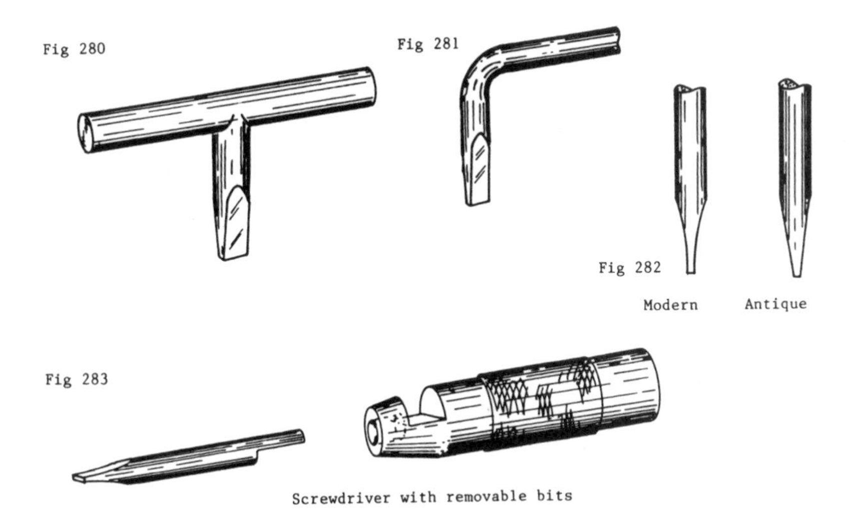

Screwdriver with removable bits

necessary, but there is no other really successful way to clean the teeth of a clock wheel, so that it is free of dirt and attractive to look at. The brass brush will also remove tarnish from brass parts without removing any measurable amount of metal. They can be bought from suppliers or as suede-cleaning brushes. The latter does not have so dense a thicket of wires as the type supplied by clock-materials companies, but I find this an advantage.

Bench

A great deal of clock repairing can be carried out at the kitchen table or on a desk, but as soon as one gets into the business of removing metal with tools, a bench is more useful.

The purpose of the bench is to present work at a workable height, to provide a firm base, and also to provide a convenient surface on which to lay out tools and materials. This may seem perfectly obvious, but few workshops are equipped for a variety of working heights, many benches are not firm, and even more do not have a depth to them that is conducive to good housekeeping. The last is probably the least thought about; benches are very commonly too wide. A radius of 450mm (18in) is a convenient 'reach' for a seated person (it will vary according to length of arm), 550mm (22in) is convenient when standing. If a bench is fastened to a wall it ought not to be deeper than 450mm; and storage shelves should be placed so that as tools and materials are finished with, they can be easily replaced, leaving the bench as a working, not a storage, space. A free-standing bench with access from both sides is a little more difficult to specify because it requires depth for stability when sawing or any other robust work. However, if you bear in mind the radii quoted for comfortable reach, you will be able to design a bench to suit your own workshop that does not waste space. Generally speaking it is not a good idea to have storage space in the centre of a free-standing bench; it prevents you cleaning the bench as often as you should, and there is a tendency for things to topple over if a heavy filing or sawing operation is taking place.

The height of working is very important. Try to provide two heights for working when standing, one at waist height, and another 200mm to 250mm (9in to 10in) lower. The first of these is handy for sawing or heavy filing; it represents the height of the vice jaws, so that the provision of two heights does not necessarily mean providing two benches, the vice can be raised by other means.

Much of the time spent at the bench will not be occupied with work that must be done standing, and a selection of different seating heights is very useful, ranging from a low stool that puts the bench at your eye-level, to a chair that is either cushioned or raised to a little over normal seating.

***Lighting**

In addition to good overhead lighting, you will need a 60 watt bench lamp with an adjustable shade. Make sure that the power socket is in a suitable place so that cables do not trail across the bench, or your feet.

***Vice**

A carpenter's vice is very useful and can be lined with hardwood for holding brass sheets and strip, but an engineer's pattern will be needed to hold large pieces firmly when hacksawing or bending. Jaws that are 150mm (6in) wide are suitable for the former, and 100mm (4in) wide for the latter. You will also find that a cheap bench-clamped vice with 50mm (2in) jaws is very useful, because it can be moved around from one height of bench to another.

***Files**

Choose a good make of file that will last for years; almost all the packaged ones sold in DIY shops are cheap, and wear out in a matter of weeks. Good, long-lasting files are made in Sheffield, by some USA tool companies and by Sandvik of Sweden. Of recent years Switzerland has manufactured Precision files, which are probably the equal of Sheffield files but tend to be more expensive. You will find it safer and, in the long run, cheaper, to buy British; the same injunction holds true for most saw-blades and twist drills.

Standard workshop practice recommends the use of new files on soft materials such as brass, and lightly used files (that have already been used on brass) for steel. New, sharp teeth cut deeply into the metal, and a hard material will put more strain on the tooth than a soft one. When the sharp edge has been dulled somewhat, and the file does not cut so fiercely, it is less likely to break a tooth when filing steel. Rubbing chalk over a file will help the swarf to drop out of the teeth. If you look at a file that has been used without dressing in this way, you will see small particles of metal jammed in between the teeth. These particles will prevent new swarf clearing away after each filing stroke, and score the surface of the metal being filed quite badly. A clean file cuts evenly over the metal so that, although it may be a rough file, all the marks are at nearly the same depth, and progressive use of finer files and emery paper will produce an unscored finish. Useful rules are:

Never lay a file down on a hard surface such as another file.

Fit handles to all files that are large enough to lie against the palm of the hand when being used.

When fitting a handle, do not hold it and stab down to drive the file-tang home. Hold the file, and tap the handle onto it.

Clean files with a file card (a special stiff brush), brushing along the teeth to clear out the swarf.

USE AND TYPE OF FILE (FIG 284)

For heavy removal of metal on clock plates, dials etc	Flat, round and half-round in two lengths 250mm or 300mm (10in or 12in) rough and second-cut
For general clockmaking work	Flat, round, half-round, square, triangular (three-square), pippin, crossing, knife of 150mm (6in) length and medium and smooth grades

Although 'flat' is a generally used term, there are types of file with flat surfaces on both sides; the most useful has parallel edges and is called a hand file. All the above files are measured along their cutting surface and are 'double cut' which gives the familiar cross-hatching effect to the cutting faces.

For removing metal from the surface of plates or parts that are to be made flush	Round-edged flat of single cut and 100mm or 150mm (4in or 6in) length, medium grade

A single-cut file is not cross-hatched in appearance, each tooth is a single cutting edge that stretches from one side of the file to the other. It does not cut as fast as a double-cut file, but it is less prone to breakage of the edge since this has more support than the pointed double-cut tooth, and the result is a reduced tendency to score the surface of the work. In particular the single-cut file is used for dressing off the ends of bushes that have been inserted into the clock plates, since it is important that the plate does not become scratched deeply in the process.

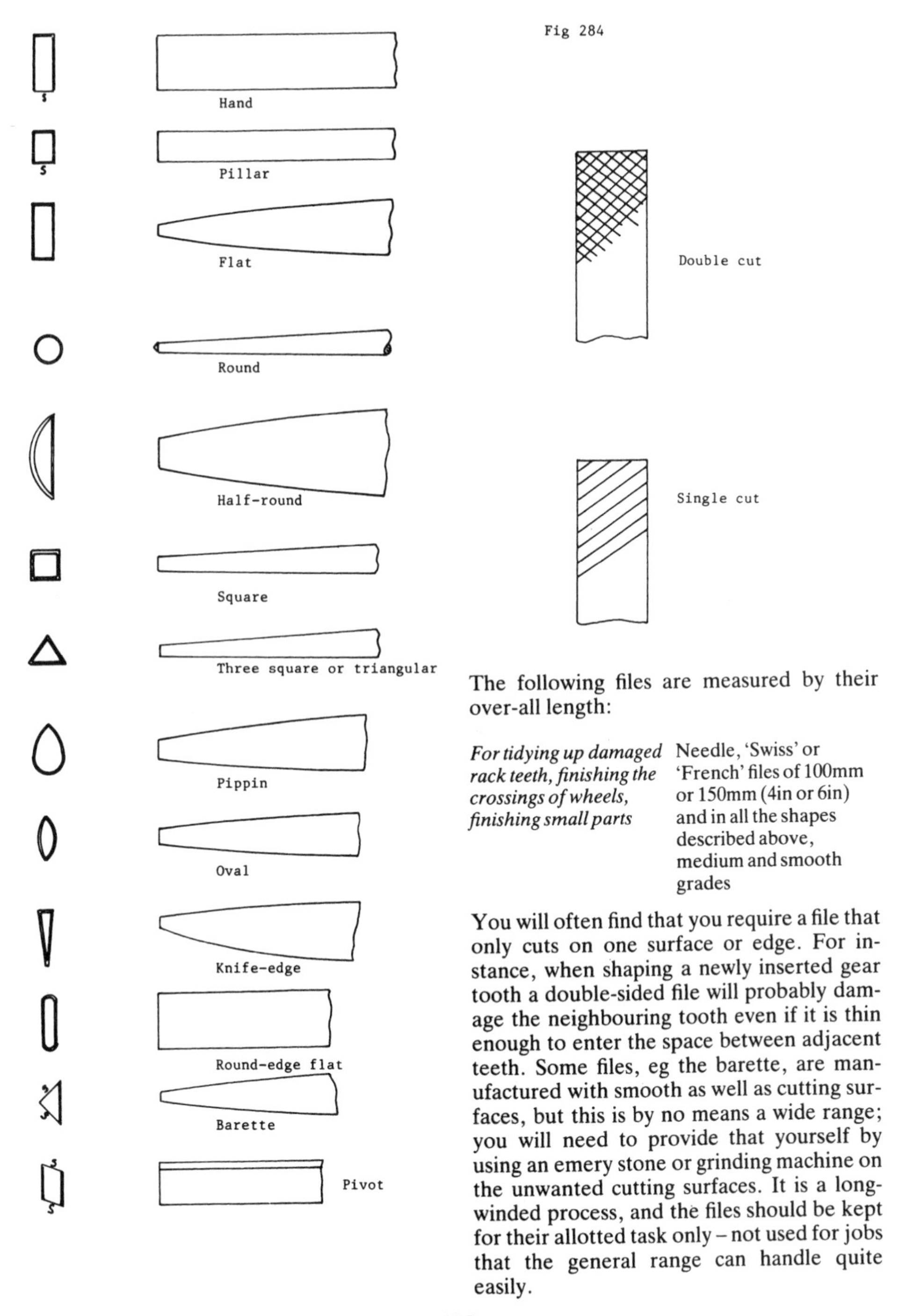

The following files are measured by their over-all length:

For tidying up damaged rack teeth, finishing the crossings of wheels, finishing small parts	Needle, 'Swiss' or 'French' files of 100mm or 150mm (4in or 6in) and in all the shapes described above, medium and smooth grades

You will often find that you require a file that only cuts on one surface or edge. For instance, when shaping a newly inserted gear tooth a double-sided file will probably damage the neighbouring tooth even if it is thin enough to enter the space between adjacent teeth. Some files, eg the barette, are manufactured with smooth as well as cutting surfaces, but this is by no means a wide range; you will need to provide that yourself by using an emery stone or grinding machine on the unwanted cutting surfaces. It is a long-winded process, and the files should be kept for their allotted task only – not used for jobs that the general range can handle quite easily.

For filing square and round holes in, eg, the gathering pallet	Miniature files, round and square 75mm (3in) long, one grade only
For cleaning up the teeth of damaged escapements, and similar small work	Escapement files have teeth on about one-quarter of their length of 140mm (5.5in), and they are made in most of the sections normal to needle files. The file is tapered along its length
For filing small localised areas or sunken places	Rifflers, the most commonly used by clockmakers being the bull's foot, but even this is not a positive requirement
Pivots of clock arbors ought to be re-surfaced whenever they are found to be rough or the plate has been bushed.	Pivot files are often provided with an angled, smooth edge so that there is no danger of producing a radius at the shoulder.
Very fine, flat files are used to produce a surface that is then burnished	You will need one of each hand (the angle creates two different-'handed' files)
Screws that are made for antique clocks ought to have tapered driver slots; in addition damaged slots of old screws can be made good again by filing.	Slotting files are best described as double-edged knife files about 50mm (2in) long, and parallel. They are quite thin, and intended for the smaller screws that a standard knife edge will not suit

MAKING A FILE

It is sometimes necessary to make a file of special cross-section, either because it does not exist as a standard product, or because you have lost or broken such a file, and cannot wait until a new one is available. Take a piece of silver steel and shape it into the cross-section required, making sure that it is fully annealed, ie heated to dull red and cooled very slowly. Now choose a double-cut file of the same number of cuts per centimetre (inch) as are needed on the new file. Hold the silver steel rigidly in the vice, and draw the file obliquely over it, the intention being to move the file in the same direction as the sloping line of its teeth. If this is done carefully it will produce a single-cut file in the softer metal, the angle of the teeth being dependent on the angle that the file is drawn across at. When successive passes of the file have formed sharp teeth on the silver steel, it must be hardened by raising to bright-red heat, and suddenly cooling in oil. An alternative for very fine cuts of file is to draw a coarse 80 grit emery paper, backed with wood or metal, over the silver steel instead of using a file. Such a tool will perform well, after hardening, on the damaged teeth of an escape wheel.

***Saws**

Two types of hacksaw – standard and junior – will be needed, the first taking blades from 250mm to 300mm (10in to 12in) in either carbon steel or high-speed steel or the so-called flexible high-speed that is constructed from both (hard teeth on a tough, flexible body). The junior hacksaw has smaller blades about 150mm (6in) long and is made in carbon steel only. The difference between the two materials – carbon steel and high-speed steel – is that although the first is slightly harder than high-speed steel, it loses its hardness very rapidly as it heats up in working. In addition, carbon steel is not so brittle as high-speed. Since junior hacksaws have carbon-steel blades by force of circumstances, you might just as well restrict your stock of standard blades to high-speed, preferably the flexible, composite type.

The teeth on the blade vary in size, and are graded by the number per centimetre or inch; the larger the number of teeth the smaller is their pitch. In order to make a choice of grade for a particular job, it is necessary to look at the effects that altering the pitch has on the cutting characteristics. Teeth with a wide spacing between them will clear the swarf produced easily, so that cutting is speeded up; however, the load on the individual teeth also increases, so that there is a greater chance of breaking them. Soft materials that put little load on the teeth can be cut with wide-pitched teeth, hard materials demand a finer pitch. In addition to this consideration of the hardness there is the fact that the teeth only cut because there is a force pressing them into the metal, and if there are too many teeth the saw will cut slowly. Sheet metal is, of course, thin compared with a bar, and the saw should be chosen so that, if possible, at least three teeth at a time are in contact with the sheet, preventing the likelihood of the teeth being broken. The saw can be angled to maintain this degree of contact. It also makes the operation less noisy.

PIERCING SAWS

These are thin, wire-like blades carried in a U-shaped frame; they are brittle and have a finer range of tooth pitch than hacksaws. The work is supported on a flat board that has a 'V' cut into it, and this board is clamped to the bench so that the gap overhangs the edge. The sheet to be cut is laid over the V and the saw is stroked up and down, cutting on the down stroke only, as the work is moved by hand over the board (Fig 285). Keep the blade as close to the edge or junction of the V as is possible, so as to support the sheet.

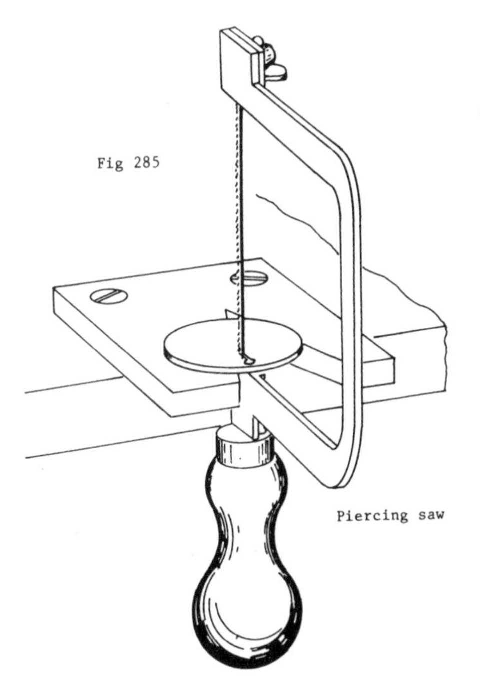

Fig 285

Typically a jeweller's or silversmith's tool, in clockmaking the saw is mainly used for crossing out the spokes of the clock wheels, or for making a hand. The blades are cheap, which is just as well because, although I can make a hacksaw blade last through three or four sharpenings, I have not yet found a piercing saw that will last more than half a dozen crossings. Professional clock-hand makers make their own blades, which are reputed to last all day.

Incidentally the sharpening of high-speed hacksaw blades is carried out with a triangular diamond file.

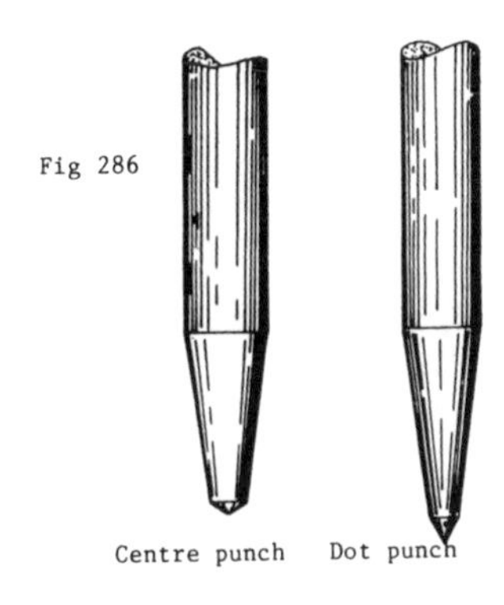

Fig 286

***Dot and centre punches**

These two types of punch look alike but serve different purposes (Fig 286). The dot punch is used to make a more permanent mark than a pencilled or scratched line; it is slim, and has an included angle of about 60 degrees so that the point can be accurately placed on a position without the shoulder obscuring the point. The mark is made with a tap from a light hammer, although I have known horny-handed markers who did it with simple hand pressure.

The centre punch, on the other hand, is used to provide a starting position for a drill or to draw a drill sideways from a false start. It has an included angle that is the same as the drill point, 120 degrees, and the shoulder tends to obscure the point so that it is best to use a dot punch first and then feel your way into position, if there is a need for accuracy.

***Hand-drill**

The traditional clockmaker's drill is the Archimedean drill which consists of a central, twisted spindle with a bobbin that slides up and down it. Moving the bobbin causes the spindle to rotate, and the drill-chuck and bit mounted on its lower end rotate also. It is an oscillatory movement, and pressure should only be put on the drill when the bobbin is moving towards the work; the return stroke runs the bit backwards. Whether the drill cuts clockwise or anti-clockwise depends on the twist of the Archimedean spiral. They are often left-hand twist, cutting on the anti-clockwise rotation, which means that a drill-bit that is used for this tool is of no use for the lathe or any other drilling machine.

It is relatively easy to use the Archimedean drill without any sideways thrust on the bit because the direction of the operating hand is along the axis. As a consequence it is less likely to break very small drills than the engineer's pattern. The latter is also called a wheelbrace, because the drill-chuck is rotated by turning a large bevel gear (to which a handle is attached) that engages a smaller bevel on the drill spindle; it is a familiar tool in the handyman's workshop. Wheelbraces are ideal for drill-bits that are between 6mm and 1mm (0.25in and 0.04in) diameter; they give continuous cutting, and the work can be carried out faster. However, because the tool is operated by turning a handle, there is a tendency to move the body of the drill at each turn and this can very easily break the smaller sizes of bit. Below 1mm (0.04in) diameter the Archimedean drill is the safest to use.

A motor-driven pillar drill will operate successfully with small and large drills as long as the spindle is steady when in use; many drillstands for pistol drills are *not* steady. A motor of between 100 watts and 200 watts will provide sufficient power for most clockmaking needs. Choose the simple kind with a straightforward lever advancement of the spindle (or quill), rather than any form of gearing, so that you can feel the cutting of the drill bit more directly, and judge whether all is well. It is much easier to break drill bits if there is a lack of 'feel' to the machine.

***Drill bits**

There are several patterns of drill bit – twist drill, spear point, flat drill and straight flute are the patterns that mainly concern clockmakers (Figs 287 to 290). For any diameter greater than 1.5mm (0.06in), I prefer twist drills of Sheffield make for the same reasons as my preference in files and saws – they usually last longer. One of my most frequently used drills, at 0.25in, must be at least twenty years old. High-speed steel is the most suitable material for drill bits, because there is a strong tendency for carbon-steel drills to lose the 'land' that runs down the leading edge of the flute where the temperature builds up and causes the steel to lose its strength. The cutting edge also increases in temperature, of course, but this is rectified when the drill is sharpened; the worn land cannot be put right.

Smaller diameters than 1.5mm (0.06in) are a little delicate as twist drills. The torque that is imposed by cutting twists the drill along its length slightly, and if the drill has twisted flutes this has the effect of increasing its length momentarily. As a result of the drill bit increasing its length, the point or cutting edges dig deeper into the metal, and there is a tendency to pull into the work, increase the strain within the drill bit, and break it off. A straight flute or a flat drill does not increase its length when subjected to the same conditions.

It is well worth collecting the now obsolete sizes of drills – letter and number sizes and the imperial sizes – because often a drill is required that is either slightly larger or smaller than the preferred metric dimension. Preferred dimensions are simply those that the tool industry is willing to stock and sell in small quantities; a purchaser of large quantities can have any size he cares to specify. Letter and number sizes, with their decimal equivalents can be found in engineers' diaries and reference books; they were often used as the tapping size for screwed holes since the optimum value rarely matched the standard fractional sizes. The following is an example of the slight variation from a nominal size that can be obtained in this way:

Nominal diameter	*Diameter in mm*
3mm	3.000
No 31	3.048
1/8in	3.175
No 30	3.264
4mm	4.000
5/32in	3.967
No 21	4.039

SPEAR-POINTED DRILLS

These are the traditional clockmaker's drill bits (Fig 288); they are easy to make, there is plenty of clearance around the shank, and they cut quite well. However, because of the clearance, a great deal of strength is sacrificed.

FLAT DRILLS

These (Fig 289) have a little more strength than the spear-point. Again, quite easy to make since they consist of a round shank, and a simple flat extension of the shank to form the drill bit.

STRAIGHT FLUTED DRILLS

The flutes (Fig 290) give a little more cross-section to these bits and, consequently, greater strength. I prefer them for the small sizes of drilled hole. All these last three types must be withdrawn frequently from the hole that is being drilled, since there is no twist to lift the swarf out and the chances of drill and swarf binding together are greater than with the twist drills.

Special drills

There are a number of drill bits that can be made in the workshop with little trouble, but which are very convenient (Fig 291).

CONE DRILLS

For opening up holes in thin material. A normal twist drill will not work satisfactorily in material that is less than the depth of its point (Fig 292), because if the tip of the bit breaks through the sheet before the lands have made contact with the drilled material, the drill bit has little stability and a misshapen hole results. Use a normal twist drill to cut an undersized, if odd-shaped, hole; then bring it to the required dimension with the cone drill. Bear in mind that the hole will be tapered unless the drill bit passes right through and onto the parallel shank.

DEE-BITS

These are exceedingly useful tools since they can be used to drill, counter-bore, chamfer, produce oil-cups, or form the decorative 'turning' that appears around the winding holes of many brass dials and the pallets of the Brocot escapement. A dee-bit is made by selecting or turning a bar of carbon steel (silver steel usually) to the diameter required and then filing a flat across the outside at one end until the end is D-shaped (Fig 293). It usually needs some sort of starting hole to follow, but it can be used without if the support spindle is rigid, or the drill is guided by a jig.

The form will vary, of course, but it can be a flat-ended bit for counter-bores, hemispherical for oil-cups, contoured for the 'turning' mentioned above, and it may have a pilot diameter turned on the end to keep the drilling concentric with a starting hole. Essentially it is a single-edged cutting tool; the choice of position for the flat that gives the cutting edges will depend on the service required. If there is plenty of support for the dee-bit, the flat can be set on the centre-line; but if the drill needs to borrow support from the surface produced as it cuts into the metal, the flat should fall short of the centre so that the D is rather more than a semicircle. Following this procedure when making an oil-cup cutting tool will ensure that when the hemispherical end has cut down a certain distance it will cease to cut further and simply burnish the cup.

A dee-bit that is turned with a slightly eccentric shank can be used in the lathe for boring out drilled holes too small for normal boring tools (Fig 294). The hardening method for this tool when using silver steel is to raise to dull red, quench in oil, and then temper to a very light straw colour. The same procedure is followed for other drills made from silver steel.

CENTRE DRILL (SLOCOMBE DRILL)

This is a short, double-ended drill bit (Fig 295) that is a combination of a straight fluted drill and a chamfering tool. It is used for making centre-holes in the end of bars that are to be supported on centre in the tailstock, and for any drilling job that calls for a drill that will not wander away from the chosen position. Since the chamfer produced is 60 degrees included angle, it will not do for countersunk screws.

***Broach**

This is a hand tool that is used to increase the size of a drilled hole to suit a pivot or some other mating part, and to improve the surface finish usually obtained from plain drilling. It is a five-sided, tapered piece of hardened carbon steel and is made in a range of sizes; the top of the taper meets a parallel cutting section about 5mm (0.2in) long, of the same diameter as the shank. The following list of sizes will meet a large number of clockmaking requirements:

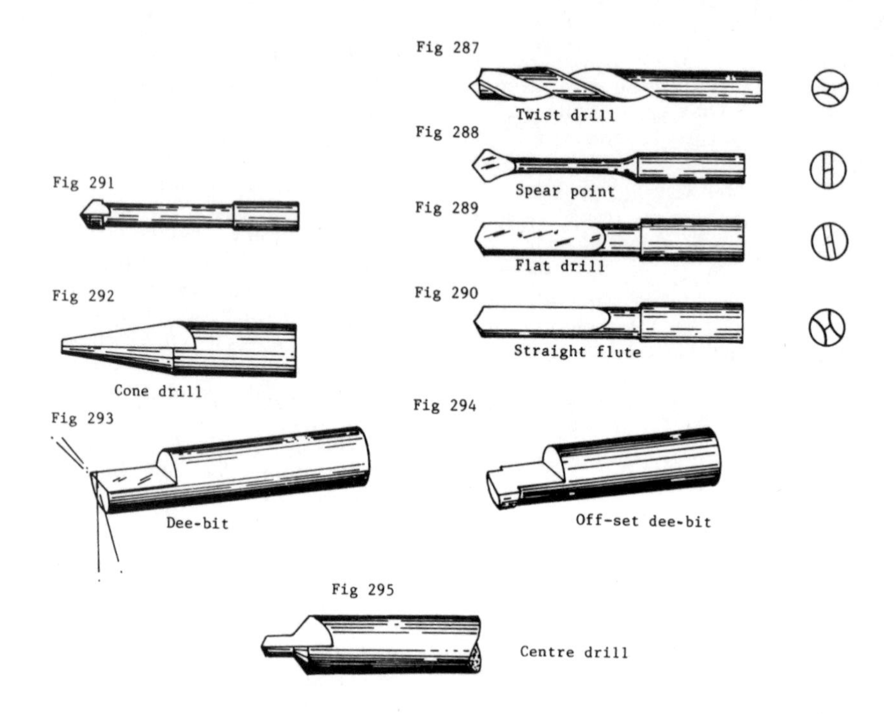

SWG	mm	in	SWG	mm	in
66	0.81	0.032	38	2.56	0.101
62	0.94	0.037	34	2.79	0.110
56	1.14	0.045	30	3.22	0.127
54	1.39	0.055	26	3.7	0.146
52	1.6	0.063	20	4.08	0.161
48	1.9	0.075	16	4.44	0.175
44	2.15	0.085	10	4.85	0.191
42	2.33	0.092	A	5.94	0.234
			F	6.52	0.257

New broaches are sometimes extremely sharp and, when used on a metal that is not free-cutting, tend to stick in the hole as they bite deeper than normal. This can lead to breakage or a very poor finish within the hole. One or two light strokes with a Arkansas stone along each cutting edge will stop it 'grabbing', and improve the finish.

POLISHING BROACH

After a hole has been drilled or broached for a pivot to run in, it should be brought to a higher degree of finish by polishing. This can be done with a polishing broach which is tapered, and given a slight cut by using fine emery paper along its length. It has a circular cross-section rather than the pentagonal form of the normal or cutting broach. Most clock repairers prefer to use this broach from both sides of the plate when producing pivot holes, so that the hole has a double taper; the smallest diameter is then at the middle of the plate thickness. The exception to this is in pre-1700 British clocks, which quite commonly have tapered pivots and the matching holes should be single taper with the wide diameter inside the clock movement.

I prefer to produce parallel polished holes, arguing that parallel pivots should not run in tapered holes, and the tool I use is shown in Fig 296. It is a simple piece of pivot steel held in a softer metal – brass or aluminium – so that it is truly square to the face of the holder. A piece of paper is interposed between the

Fig 296

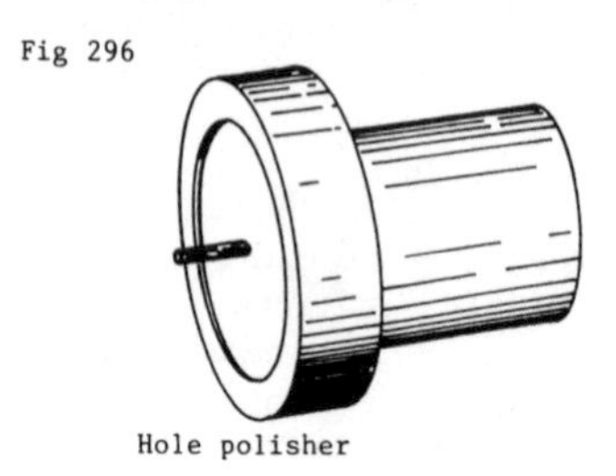

Hole polisher

holder and the clock plate, and then the tool is rubbed around the inside of the hole by rotating the whole thing – not twisting it upon its axis, but causing it to drag on the inside of the hole, burnishing it.

It is claimed that polishing broaches produce a work-hardened surface; but since it needs about 25 per cent of deformation to appreciably work-harden soft 70/30 brass, the diameter of the hole would have to be increased by something approaching this for there to be any real effect other than simple burnishing.

Steel ruler

Very little measurement in clock repairing is made with a ruler, but it is a cheap way of obtaining a straight-edge, which is used for drawing marking-out and construction lines. Since the ruler is to be used mainly for this purpose, pay more attention to the condition of the edge than the dimension marks. The straightness can be tested by putting two rulers edge to edge, and then viewing a light with the rulers between you and the light source. Any space of more than 0.025mm (0.001in) will show light quite clearly. A ruler of 250mm (10in) length and about 1mm (0.04in) thickness will do very nicely.

Scriber

This is a simple piece of hardened carbon steel – pivot steel is ideal – about 2mm (0.08in) diameter ground to a point with an included angle of about 15 degrees. It is used for making construction marks on brass or steel, but never on surfaces that are required to be scratch-free, since a scribed line can easily be 0.125mm (0.005in) deep and this would be difficult to remove afterwards.

***Emery paper and other abrasives**

Abrasives are used to remove tiny amounts of metal from soft materials, or to remove any amount from a metal too hard to cut with tools, or to polish the metal.

EMERY PAPER

This will be needed in a wide range of grit sizes – metal finishing depends upon it. The grit size governs the coarseness of the cut that is made by the paper; if a grit is large it will make a deep scratch in the metal that you use it on, and if it is small the scratches will be much shallower. The measurement of the grit size is made by sieving the abrasive during manufacture, and the size is stated by quoting the number of holes per square inch in the sieve – I do not believe that this has altered with metrication – which means that small numbers denote large grits, and large numbers small or fine grits. A great deal of careful grading goes into the manufacture of a good emery paper, to ensure that all the grits are of similar size, and that there are no oversized grits that will make deep scores in the abraded surface; all this care will be wasted if the paper is allowed to pick up dirt and grit from the bench. Take care when using the emery paper that nothing falls onto it, and stays there. Rap the paper on something clean, such as the leg of the bench, to dislodge any grit and, of course, metal that it has already removed from the work.

A useful selection of grit sizes follows, grading from coarse to fine, which is the order in which emery paper is used so that the finish becomes progressively finer. Do not get too many sheets, two or three will suffice, replacing as necessary. I prefer to stock wet and dry paper, because the polishing of brass dials and the removal of any excess black wax from restoring the numerals requires wet emery paper.

Sizes: 40, 80, 120, 200, 400, 1000 and 'crocus' (an extremely fine abrasive that is a peculiar red/blue colour).

METHOD OF USE

All polishing of metal should be done in one direction only, do not alter the direction that the paper or stone is stroked or you will produce a cross-hatching of lines that makes final polishing very long-winded. Wherever possible the strokes should be made along the length of the workpiece, but this will depend on its shape, and any obstructions to a clear stroke.

Paper should be backed with some firm material when polishing plane surfaces, particularly if there are holes in the surface as in the case of clock plates. If you simply use fingers for backing the paper, the edges of the holes will become rounded and blurred; only use fingers if it is necessary to clean a hollow or bit of local damage. Pieces of blockboard, planed timber or brass are useful for backing the paper when dealing with simple flat surfaces. 'Oasis' – a flower arranger's material – is ideal for backing when a curve has to be accommodated. It readily conforms to any manner of contour, but be careful at edges or holes for it is not sufficiently firm to maintain a crisp edge. Whether the emery paper is stroked along the work or the work is stroked over the abrasive is a matter for the particular task; I find the polishing of pieces such as racks and rack hooks easier if the paper is mounted on a flat board, and the work is stroked over it. It is easier to obtain a flat surface by pinning a sheet of paper to a board than by gluing it; the glue tends to wrinkle the paper. Pin by turning the edges of the paper under the board so that the fastening does not impede the progress of the work.

EMERY STONES AND INDIA STONES

Commonly called slipstones, these are mainly used for localised work on hard materials – re-surfacing escape pallets, sharpening tools etc. The most useful is a simple rectangular block about 75mm x 25mm x 8mm (3in x 1in x 0.315in), medium grade. Both are used for sharpening tools and will produce a fine edge that can be brought to an even higher finish by the use of an Arkansas stone.

ARKANSAS STONE

A grey/white hard-stone, very fine abrasive that is used for putting a final edge on tools and is particularly useful for polishing the surfaces of escapement pallets. It is supplied as small sticks about 75mm (3in) long in a range of sections. A square, triangular and round stick should meet all normal requirements.

WATER OF AYR

This is used by many people for finishing brass; it is slate-like and is used wet. I prefer to use wet and dry emery paper on a board.

***Gas torch**

The modern gas torch running on butane or propane makes the jobs of soldering, heat treatment etc very much simpler than the old clock repairer's charcoal, blowpipe and spirit lamp did. Whether you use a composite gas torch/gas container such as the familiar handyman's torch, or a torch, hose and large floor-standing container, will make no difference to the size of work that can be heated. The size of the nozzle and jet governs the amount of heat that can be put into the work, the other considerations are just a matter of convenience and expense. Most work can be done with a nozzle about 16mm (0.625in) diameter and if a 'Cyclonic' nozzle is used (internal channels give a rotation to the flame) you will find that both narrow and wide flames can be obtained by changing the gas pressure. This is quite enough variation for coping with work as small as re-facing a recoil escapement or as large as silver-soldering 16mm (0.625in) square-section brass.

If a hose is your choice, make sure that it has a valve in it that prevents gas flow in the event of the hose being cut or catching fire, and does not allow gas to flow in the other direction – into the cylinder.

***Micrometer and vernier calliper**

Much clock repairing can be done without making any measurement of dimensions at all, by simply making parts fit each other; however, when you expand your workshop to include a lathe it is almost impossible to manage without either a micrometer, or a vernier. The micrometer should be a 0 to 25mm (0 to 1in), and it need only be a simple instrument. There is no need for special insulating pads to protect it from the heat of the hand, or vernier scales, or digital read-out. If the screw turns smoothly, the anvils (the ends of the measuring spindles) come together without light being visible between them, and the instrument is pleasant to handle, it is very probably a good instrument. Do not worry if the micrometer does not read precisely zero when the anvils touch, this can be easily adjusted. Your micrometer will be used for measuring the diameters of pivots, arbors, pinions and other turned parts; accuracy to 0.025mm (0.001in) is the most that is needed.

The vernier bears the name of the inventor of the scale, in clock repairing it is used to refer to the complete instrument although strictly speaking it is the system of measuring small dimensions that is 'vernier', not the instrument. Choose one with both metric and imperial measurements for preference; you will find it a very convenient conversion instrument for changing either set of units into the other. It is not usual to need more than

150mm (6in) of scale length, and a larger vernier is not so easy to manipulate. Its virtue lies in the fact that it will measure distances from 0 to 150mm (0 to 6in) to within ± 0.05mm (0.002in); it is more comprehensive than the micrometer, but not as reliably accurate. Many of these instruments are designed so that they can measure the inside diameter, and also the depth, of holes.

Lathe

If you are not familiar with lathes, get advice before buying one. Most repair work can be carried out on a small centre-lathe equipped with three-jaw chuck, drilling chuck and having a wide range of speeds (200–2,500rpm). It is not necessary to look for a lathe that will swing more than 75mm (3in) over the bed or turn more than 250mm (10in) between centres); you could manage with less. A tough lathe that can take heavy cuts will save a great deal of time compared with a light lathe that has to be treated gently.

Taps and dies

It is very useful to be able to make replacement screws, and new tapped holes in the place of old mangled ones. Taps and dies are expensive to buy as full sets, so buy them as needed. The metric screw threads that seem to be favoured by clockmakers are largely the non-preferred sizes – 1.3mm, 1.7mm and so on – which means that it is not always easy to obtain replacement screws at short notice. BA threads are also metric diameters, and come close to many of the screws commonly found in clocks; the screws can be bought from hardware shops rather than specialist suppliers, and modified to match the other screws in the clock.

BA	2	4	6	8	10	12	14
Dia	4.7	3.6	2.8	2.2	1.7	1.3	1.0mm

Unless there is a likelihood of your having to make a great many screws and screwed holes, carbon steel is quite satisfactory for hand-held taps and dies. Choose a reputable make, Sheffield for instance, or you will find that the screws cut in the dies will not suit the holes that are cut with the taps.

Soldering hearth

As the complexity of the work that you solder, silver solder or heat treat increases, you will find that it is more convenient to have a properly protected part of the bench for this work, than simply carry it out on the odd piece of sheet metal.

A hearth can be made very cheaply from ordinary house bricks (dry them out in a slow oven first), the internal blocks of an old storage heater or fire-bricks. The two last are the best. Choose a well-ventilated part of the workshop, and a waist-high bench or shelf, and cover the latter with sheet metal for the extent of the intended hearth to prevent charring as a result of hot gases being forced between the bricks. The hearth need only be a flat layer of brick with a wall of brick on three sides. A spare brick that can be laid over the top as a roof will be useful when dealing with large pieces of metal; it contains the heat which otherwise would escape (Fig 297).

Fig 297

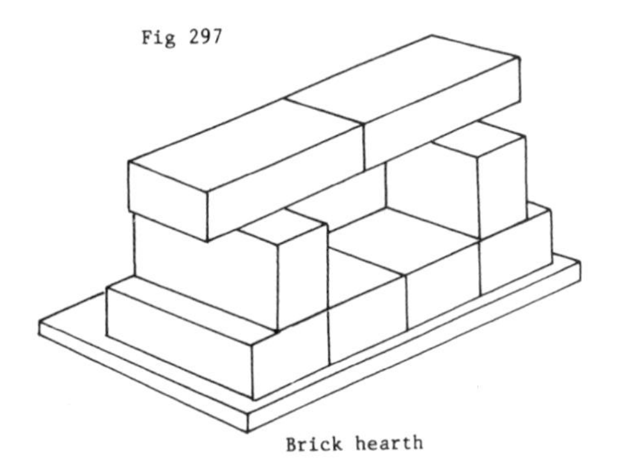

Brick hearth

Depthing tool

Buy one that will take the longest arbors and largest wheels that you are likely to come across. If possible, get a free-standing one with a positive lock.

Tin snips

Duck-billed snips with 250mm (10in) handles are needed for hard work, and a smaller pair with 125mm (5in) handles for lighter work. The former are thicker in cross-section than the ordinary snips, and not so high from cutting edge to back. This gives them strength and still enables them to work along a piece of sheet metal without the back of the snips fouling the cut-off piece of metal. Snips can be used most easily if one handle is gripped in the vice, and the sheet is brought to the snips.

Planishing hammer

In addition to an ordinary hammer for use with centre-punches etc, you ought to have a hammer that is used for nothing else but copper alloys. What type it is does not matter so much as that it should have a polished flat face. This tool is for flattening brass parts, finishing off riveted pillars or tidying up the damaged edges of brass plates; since the hammer is polished and the face is flat with no dents, it is possible to use it on brass without damaging it. It is a good idea to paint the handle an outstanding colour so that it will not be picked up in error and used for unsuitable work.

Special tools

Here are one or two tools that can be made as needed, and which may solve an awkward problem or two.

CENTRING PUNCH

This is for marking the centre of rods for centre-drilling when the work is too long for the lathe with drill chuck (Fig 298).

HOLLOW MILL

If the centre arbor of a clock is broken off at the end, you can finish up with a chain of events that require that the dial be brought closer to the clock movement. This can be done by making a tool that will machine the shoulders of the dial pillars – the hollow mill is such a tool. There are a few more repairs that are carried out more easily if the shoulders of pillars can be faced back in this manner.

CHAIN-DRILLING PUNCH

Marked-out shapes in sheet metal are often cut out by drilling a series of holes around the marking, and then sawing or chiselling through the space between the holes. The system works best if the holes are very close to each other and, if two punches are placed in the same body, centre dots at predetermined centres can be produced rapidly (Fig 299).

THIN-SHEET BROACH

Large holes in thin sheet are difficult to open up with normal broaches because they tend to produce non-circular holes when large diameters are involved. A broach with only one or two cutting edges will not produce an irregular hole because of the support of the non-cutting part of the broach (Fig 300).

Fig 298

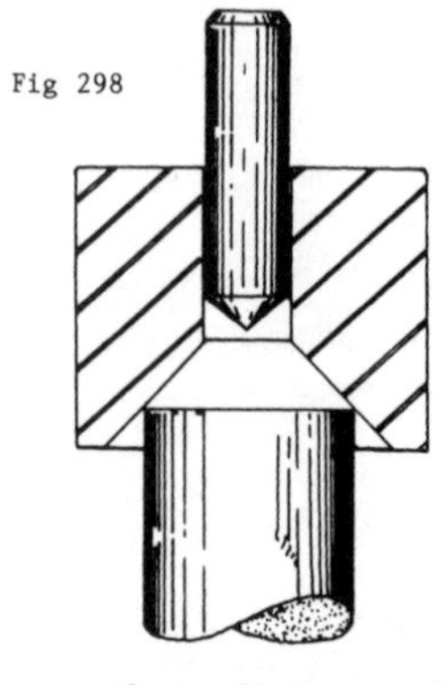

Centre finding punch

Fig 299

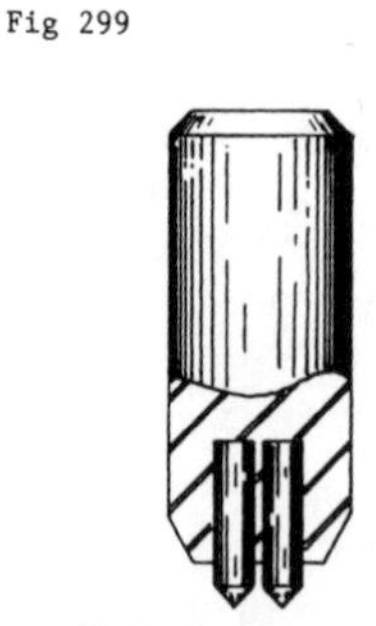

Chain-drilling centre punch

Fig 300

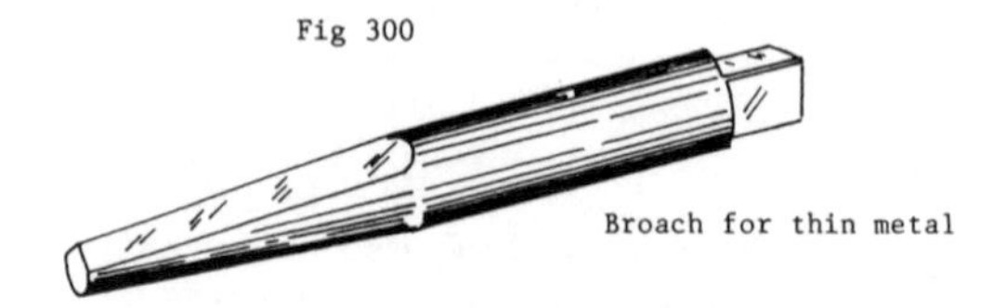

Broach for thin metal

Materials needed

Few materials are needed for clock repairing. One can, of course, purchase all manner of parts from suppliers – castings, dials, springs, screws etc – but a catalogue will be of more use than additions to this list. The materials that ought to be kept in your workshop stock are brass and steel, which are the raw materials common to almost every type of clock movement.

BRASS

You will require this in two forms, sheet and rod, and the latter only if you are intending to carry out turning operations. Brass is not cheap, particularly in small quantities, so restrict the stock to relatively thin sheet of less than 1.5mm (0.06in) for instance. Thicker pieces needed for repairing wheels, cocks, plates etc may be ordered as required from a non-ferrous stockholder. Brass rod is still available in imperial dimensions as well as metric and this, too, ought to be ordered as needed. In both cases – sheet and rod – the quantities required are usually so small that there is little difficulty in obtaining them by post if a stockholder is not available in your area. Different types of brass are listed below with their major characteristics and uses; the numbers given will be recognised by stockholders.

CZ108 A 'half-hard' brass of 70 per cent copper and 30 per cent zinc, which can be worked when cold; it is therefore used for parts that have to be bent or beaten into shape. However, it is not very easy to tap because the metal tends to cling to the tool during the threading operation. Most antique clock plates were made from 70/30, probably with the addition of lead, and if it is required that bushes or other parts let into old clock plates match the colour of the plate, 70/30 must be used despite its poor tapping quality. Cast rods of 70/30 are available from some suppliers for making new pillars and, as just mentioned, bushes.

CZ119 and CZ121 Machining-brass that is supplied in rod or extruded section form. It can be worked in the red-hot condition, but does not respond well to cold working, and will crack if much bending or beating is done in this condition. It approximates to the proportions 60 per cent copper 40 per cent zinc, but it also contains a small amount of lead to give it good machining qualities. It can be tapped, turned and drilled easily, but it has a slightly different colour to 70/30 which becomes obvious when it is inserted as bushing material. This alloy is useful for machined parts, fabricated parts that are silver soldered together, and any structural use that does not require cold working. It is also useful for gear wheels.

CZ120 This is often called engraving brass, and its most useful form is highly polished plate or sheet that can be used for making new clock plates. It machines very easily, and can also be used for gear wheels.

STEEL

Silver steel is a high-carbon steel that is easily bought from tool shops and model shops as well as from ferrous stockholders. High-carbon steel can be heat treated to make cutting tools, springs, pinions and other clock parts subject to hard wearing conditions. It can also be used in the unhardened state so that it is often unnecessary to stock the softer mild steel, which makes things simpler in the workshop. Mild steel cannot be hardened by direct heat treatment, and if both types are kept in the workshop it is difficult to recognise which is which. Silver steel is supplied in lengths that are quoted in imperial measure 13in and 39in (approximately 330mm and 1,000mm).

Another source of high-carbon steel is pivot steel, a material that is supplied in lengths of about 100mm (4in) in a wide range of diameters. It comes hardened, tempered blue and finished with a high polish. Although its major use is the replacement of pivots by drilling the arbor and insertion of the pivot, it is a very handy material for springs, small screwdrivers, scribers and any number of other small implements. It is well worthwhile obtaining a selection of sizes; most suppliers carry selections as well as packets of single sizes.

Supports

At all stages in the development of your workshop, you will need to support the clock movement in a manner that allows you to test it, and possibly carry out adjustments. Some

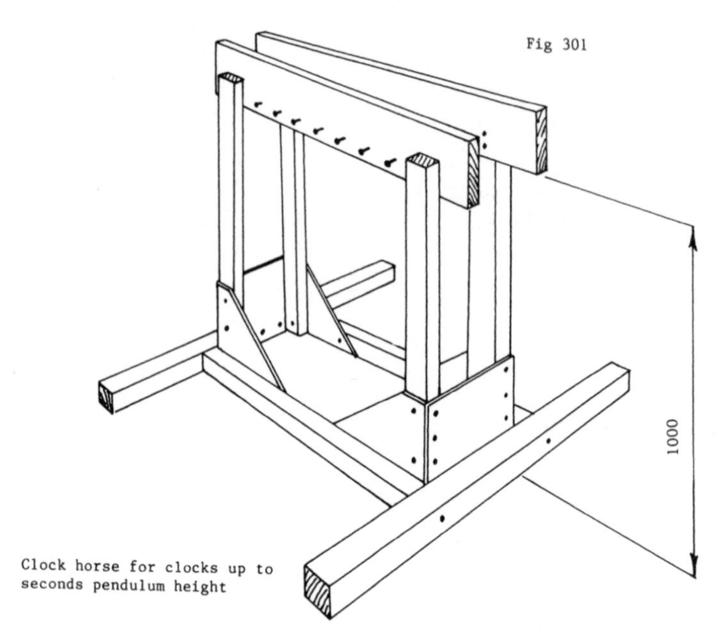

Clock horse for clocks up to seconds pendulum height

movements have a flat base and can stand directly on the bench or, if there is a pendulum hanging below the level of the movement base, on a shelf with the pendulum overhanging. This will only suit spring-driven clocks, of course; if a weight-driven clock is to be tested whilst standing on a shelf, holes will need to be provided in the shelf for the weight cords.

Since most weight-driven clocks have a seat board to which the movement is bolted, you may find that a couple of L-shaped brackets fastened to the wall and spaced to suit the seat boards will meet all your requirements.

Supports for clocks on the bench whilst they are being adjusted or assembled can be whatever comes most readily to hand – square sections of metal placed either side of round movements, rectangular wooden blocks or boxes to raise flat-based movements high enough for the pendulum to clear the bench, open-topped-tins or ends of tube that the movement can nestle in.

CLOCK HORSE

Although this is an implement that need never be used – as we have seen, a shelf or pair of brackets are sufficient for testing – it does place the movement at a height convenient for working on, and allows you to gain access to all sides of the movement. It does need plenty of floor space though, so that there is no danger of knocking against it when a clock is mounted on it; and if you cannot allow a space in your workshop of about 1m (39in) square for the horse it is better to manage without one.

Fig 301 shows the main features of a horse. It consists of two boards on edge for the clock movement to stand on, and legs to raise the boards high enough for a seconds pendulum to swing freely. A seconds pendulum is the common length found in longcase clocks; it beats once every second and its length, from suspension point to centre of bob, is about 1m (39in). To make supporting a variety of clocks a simple matter, I suggest that the two boards are not parallel to one another but are 150mm (6in) apart at one end and 75mm (3in) at the other. A row of short nails along both boards will provide convenient anchorages for the ends of the weight cords. When setting a movement in place on the horse follow the same procedure as laid out in Chapter 1, ie weights are attached first and taken off last. Beware of spring-driven clocks that have the barrel or a wheel protruding below the base, they will 'walk' along the bench if they are not lifted up on supports.

Glossary

Anchor Anchor pallets is the common term for the recoil pallets of a domestic clock such as longcase, bracket and dial clocks. *See* Escapement, recoil. Strictly speaking, it refers to the shape of the pallets and consequently can be applied to some dead-beat escapements.

The anchorage of either end of a mainspring is traditionally called a hook; the fastening into the barrel is a barrel hook, barrel stud or stud anchor.

Arbor The horological term for any spindle or shaft, eg, centre arbor, winding arbor, escape-wheel arbor.

Barrel A tube that contains the mainspring, usually fitted with a bearing at each end, one in an integral base and the other in a removable cap. Barrels intended for use with a fusee are plain cylinders; barrels that carry gear teeth at one end (barrel ring), are called going barrels, and do not work with a fusee.

Beat The swing of a pendulum from one side to the other, or the swing of a balance wheel in one direction. Two beats are needed for a complete oscillation of either pendulum or balance wheel.

Birdcage Description applied to a posted movement consisting of a horizontal plate of metal at the top and bottom joined by vertical posts. The pivot bearings are drilled into vertical strips that may be straight or cruciform. Typically, the movement is of thirty-hour duration, and weight driven by rope or chain. Striking is by count plate, or occasionally internal rack.

Bob The name given to the main mass at the end of a pendulum.

Bracket clock Developed with the application of the pendulum to British clocks *c*1658. The movement is plated, ie, it comprises plates with bearing or pivot holes and horizontal pillars joining them. Early bracket clocks had verge escapements, a practice which continued until the end of the seventeenth century. The recoil anchor escapement superseded the verge. Originally it was a portable clock, often fitted with a repeater device. Many *were* placed on brackets, but this has never been universal. Design and size vary with date and maker, but they are basically rectangular clocks with variously shaped tops and embellishments. The name is usually applied to designs developed before the mid-nineteenth century.

Break arch A semicircular piece of the dial, surmounting the main rectangular part, but not as wide, so that shoulders are left on both sides. The term is also applied to the hood that follows this form.

Broach Horological tool for increasing the size of holes – a piece of hard steel, tapered and having a polygonal (most often five-sided) cross-section. It is a form of reamer and cuts by being rotated in the hole.

Burnishing Method of polishing metal by friction with another highly polished surface. A finely abraded surface is used for burnishing pivots (a hard steel blade is rubbed with fine emery paper).

Bush Pivot holes that have become enlarged or mis-shapen can be corrected by enlarging them and then inserting a short tube with the correct bore. This tube is called a bush. When the bush is to be held by friction alone it should be of the same or slightly larger diameter as the hole that accommodates it.

Cannon pinion The gear that is mounted on the tube that carries the minute hand (called minute pipe). This gear meshes with another to drive the hour hand in normal domestic clocks; the second gear is called a minute wheel and often carries a peg that initiates the striking of a clock.

Centre wheel *See* Wheels.

Chapter ring The circle of numerals on a clock dial, with the minute and, sometimes, quarter-hour divisions. An imposed chapter ring is a separate piece of metal attached to the dial plate, usually accompanied by imposed spandrels in cast brass. *See* Spandrels.
Circular Error The rate of almost all pendulum clocks varies with arc because the pendulum arc approximates to a circular path. If it followed a cycloidal path there would be no change of rate (isochronous). The difference in rate is called circular error.
Click Horological term for a pawl, which allows a ratchet wheel to move in one direction only, eg, when winding.
Cock Support for a clock part: back cock supports the pendulum; minute-wheel cock supports the minute wheel.
Collet, hand Small brass disc fitted into a hand and onto the corresponding arbor or pipe. The position of a minute hand may be adjusted against the numerals on the dial by rotating the collet in the main body of the hand.
Collet, wheel Short brass cylinder that is soldered or otherwise fixed to an arbor, then turned and the wheel riveted in place. *See* Swage. Early clocks had iron collets 'of a piece' with the arbor.
Contrate wheel *See* Wheels
Crossing out Clock wheels have metal removed to leave spokes. This is called crossing out and serves to reduce the mass of the wheels, thereby increasing the speed of response of the gear train to the escapement or release mechanism of the strike.
Crown wheel *See* Wheels
Crutch The impulse of the escapement is transmitted to the pendulum in most clocks by a lever attached to the pallet arbor. The lever engages the pendulum rod either in the form of a fork or loop about the rod, or in the form of a short rod that lies in a slot in the rod (*See* Fig 192).
Depthing The process of achieving the best meshing of wheel and pinion.
Depthing tool Supports a meshing pair of gears between centres that are adjustable. The depth of engagement is therefore capable of being altered to obtain the best condition, and points on the adjustable centres can be used to mark the best positions for the arbor pivots on the clock plates. This process is called planting the train.
Drop-off point The position at which escape-wheel teeth leave the impulse face of escape pallets. The wheel then moves quickly through a small angle until the next tooth strikes either the impulse face (recoil escapement) or the locking face of the pallet (dead-beat), making the familiar tick.
Escape wheel *See* Wheels.
Escapement The device that releases the going train in a controlled manner so that the hands can turn. Its rate of operation is governed by an oscillating system such as a pendulum, balance wheel and spring, foliot etc.
Dead-beat escapement: type of escapement that does not result in recoil, the retrograde movement of the escape wheel at some time during the beat. The most common are Graham and Brocot.
Platform escapement: where the escapement is mounted on a separate plate to the clock plates (carriage clocks, ships' clocks). There are two common patterns: the cylinder escapement, in which the pallets are formed from a hard steel cylinder mounted concentrically with the spindle of the balance wheel (the staff); and the lever escapement, in which a forked lever carrying dead-beat pallets is oscillated by the action of the balance wheel to release the escape wheel. The latter is generally reckoned to be the better escapement and it was fitted to more expensive clocks than the cylinder escapement.
Recoil escapement: the common escapement fitted to longcase clocks and others. The escape-wheel teeth drop directly onto the impulse face of the pallets, with the result that the wheel is turned backwards against the drive of the train by the swing of the pendulum or balance during the supplementary arc. The most common recoil escapement is the anchor.
Verge escapement: The earliest type of escapement, also a recoil. It consists of a crown wheel that acts upon the impulse faces of an arbor lying parallel to the face of the wheel, these faces being small planes protruding from the arbor at opposed angles to the wheel face and on either side of its diameter. The verge may be associated with a pendulum, or (in earlier models) a balance wheel (without balance spring), or foliot. It is not normally a good timekeeper, but there is a rare version that is dead-beat rather than recoil.

Flat The rectangular sectioned piece of metal (usually brass) at the top of a pendulum rod, which is engaged by the crutch; and a similar piece at the bottom of the rod for the bob to slide on when adjusting the rate.

Fly A piece of flat brass mounted on an arbor at the top of a gear train and intended to act as an airbrake. It is associated with strike and chime trains. In a turret clock the fly is also used on occasion as a flywheel, storing energy for the operation of the striking hammers and thus allowing the train to run on a lighter weight.

Foliot An old type of oscillator, performing the same function as a pendulum but more erratically. It consists of a bar mounted at right angles to the verge pallet arbor or staff. It may have weights on each arm to adjust its rate of rotation.

Frontwork Everything mounted on the front plate of a clock except the dial.

Fusee A device to maintain an even torque on the clock train when driven by a spring barrel. It is of varying diameter according to a calculated curve, so that if a cord is wrapped around it and around the spring barrel, the torque arm of the fusee increases as the available force from the spring declines.

Gathering pallet A finger or rotating pin that engages one tooth of a rack (in some American clocks, a count wheel) so that the clock makes one hammer strike at each gathered tooth. The strike in some clocks consists of more than one stroke.

Hoop wheel *See* Wheels.

Impulse The small amount of energy that is passed to the oscillator by the action of the escape wheel on the impulse face of the pallets.

Impulse pin The pin made of hard steel or a jewel which is carried by the balance wheel and is impulsed by the lever in a lever escapement.

In-beat This is when the oscillator (pendulum etc) makes two equal semi-arcs (one in each direction), each releasing the escapement.

Intermediate wheel *See* Wheels.

Jumper A bent lever that lodges between the points of a star wheel and locates it accurately to each station or point. It is spring loaded.

Knife-edge suspension A simple fulcrum, either knife-like in cross-section or consisting of one curve resting on another and having point contact only. Used in verge movements and Black Forest clocks and some others.

Lantern clock Early British clock design developed in the first half of the seventeenth century. It is a posted clock, all metal, with a bell mounted over the top and often flat embellishments surrounding the bell and thin brass side doors on the movement. Originally weight driven, Victorian models and twentieth-century types are often spring driven.

Lantern pinion A small gear (pinion) made from a round brass body that carries a circle of steel bars or trundles. The trundles form a cylindrical cage and are spaced so as to mesh properly with a wheel.

Lifting piece The lever that is raised (by a cam on the centre arbor or a pin on the minute wheel), to initiate the striking sequence.

Line of centres A line drawn between the centres of two meshing gears. Proper meshing is obtained when gear teeth do not touch until they have reached or slightly passed the line of centres, as they rotate.

Locking A dead-beat escapement has a face on the pallets that holds the escape wheel tooth still until it is time for the impulse. This is the locking face and the amount of lock (often measured in degrees) is the distance between the point of first contact on the locking face and the beginning of the impulse. If there is no lock the result is recoiling of the escape wheel.

Maintaining power A small supply of energy that supplies motion to the escapement whilst the clock is being wound up. Going-barrel clocks and weight-driven clocks with the Huygens' loop chain do not need maintaining power. A Huygens' loop is a cord or chain that passes over two pulleys, one driving the going train and the other driving the striking train. The cord is looped so that when a weight is hung from it the resulting tension turns the train in the required direction; the rest of the cord hangs on the other side of the pulleys and when pulled down raises the weight to wind the clock. Only one pulley has a ratchet (the striking side), so that only this side is unpowered when winding.

Bolt and shutter maintaining power: An early method employing a bolt that can be pulled back against a spring, which, when released, presses a finger to turn the great wheel. The bolt carries two discs that block the entrance to the winding arbors until the bolt is withdrawn.

Gravity bar maintaining power: On tower clocks, a weighted bar carries a ratcheted finger that rests on a convenient train wheel. In the lowered position the finger does not touch the wheel; in the raised condition it engages the teeth and drives the train by its weight. A shutter is attached to block the winding arbor when lowered.
Harrison's maintaining power: A ratchet wheel is interposed between the great wheel and the barrel or fusee. This ratchet wheel carries the normal click which drives the normal winding ratchet on the barrel or fusee. Contact between the first ratchet and the great wheel is maintained only by a spring. When the second ratchet is being driven by the barrel or fusee the motion is transmitted to the first by the normal click. This ratchet in turn transmits energy to the great wheel by the pressure on the spring. A pawl on an arbor borne in the clock plates prevents the ratchet turning backwards. At any time there is a driving force maintained by the spring between the first ratchet and the great wheel, quite independent of any winding being applied. There are other types of maintaining power, but these are the most common.

Matting A treatment of the dial surface on a clock face, normally restricted to the area within the chapter ring. The best type (and the most expensive) leaves a totally random, finely 'wormed' finish that shows no pattern at all and is evenly distributed. Methods using knurling tools or other specialised tools usually leave a discernible pattern that is visible from certain angles.

Mesh Engagement of one gear with another.

Oil-cup or sink A part-spherical indentation around a pivot hole to prevent the oil spreading over the face of the plate and retain more oil at the entrance to the hole. Introduced generally in the mid-eighteenth century.

Pallets The teeth of the escape wheel or crown wheel bear on the pallets during operation of the escapement.

Pinion Horological term for a small gear, frequently of hardened and tempered steel.

Pipe Any tube in a clock. The tube that carries the hour hand is an hour pipe.

Pitch The distance between the tips of teeth or the centres of holes, normally used to refer to a dimension common to a series of teeth or hole centres.

Pitch circle In gears, this is a diameter on each gear where the teeth are considered to make first contact. It is also the diameter at which the tooth pitch should be measured. Half the sum of the pitch circles (or pitch diameters) of two properly meshing gears defines the distance between their centres. The ratio between the number of teeth on each of a meshing pair should be the same as the ratio of their pitch circles. In the Module system, the pitch circle (pitch diameter) equals the number of teeth multiplied by the Module. This is the definition of the Module which in horology has metric dimension.

Pivot The part of an arbor that turns in the bearing hole.

Pivot hole The bearing hole for a pivot.

Plated movement A clock movement that comprises two or more flat plates that provide the housing for pivot holes, pillars to separate and support the plates, and one or more trains of gears with associated work, eg, longcase, carriage, table clocks etc.

Posted movement A clock movement that comprises two or more flat plates that do *not* provide vertical housing for pivot holes etc (usually vertical). The pivot holes are contained in strips of metal or wood, eg, lantern, birdcage, cuckoo clocks etc. The plates are joined by the posts and support the strips.

Rack A toothed sector of a circle (usually associated with a snail) that forms part of the striking work. It may carry either ratchet-type teeth or gear teeth according to the operation of the clock.

Rack hook A pawl that rests on the ratchet-type rack.

Rack tail An arm attached to the rack, which engages the snail, so that the movement of the rack is governed by the dimensions of the snail at the point of contact.

Rate The amount that a clock loses or gains time. Changes of rate can be effected by altering the effective length of a pendulum, or altering the effective length of a balance spring, or altering the radius of gyration of a torsion pendulum, balance or foliot.

Ream To make a hole larger by removing metal from the bore. *See* Broach.

Repeater A clock that is made to strike the last hour sounded by depressing a button or lever, or pulling a cord.

Semi-arc The arc from the at-rest position of a pendulum or balance wheel to its furthest swing, ie, half the total swing. When both (equal) semi-arcs are the minimum needed to release the escapement, the clock is said to be in-beat.

Shake The amount of clearance in the fit of pivot, pivot hole, distance over the shoulders of the pivots, and distance between the bearings. Literally, the amount that the parts can shake. Side shake is the movement of the pivot from side to side within the pivot hole. End shake is the shake between the clock plates allowed by the space between the shoulders of the pivot and the internal face of the plate.

Shelf clock An American mass-produced clock that sits on a shelf. Generally larger than a mantel clock, it can be wagon spring, open spring or weight driven, thirty hour or eight day.

Silvering A simple process of laying a thin deposit of silver onto a copper alloy, usually brass, to give a white background on dials so that the numerals will show more clearly. It was not originally intended to be shiny, but a bright matt surface that would catch the light without high-spots. A moistened mixture of silver chloride, salt and cream of tartar (in ratio 1:1:2) is rubbed onto a clean brass surface. It is washed off and then rubbed with plain cream of tartar (moistened) to fix it. The plate is washed again and neutralised with bicarbonate of soda before final washing.

Snail A snail-shaped cam that has a curve or series of twelve flats, incremented over eleven steps to set the movement of the rack to correspond with the hour that is to be struck. The twelfth step is between twelve and one o'clock.

Spandrel A decorative feature to fill open space on a dial. They appear as castings, engravings or painting between the corners of the dial and the chapter ring, and in either side of the arch in a break-arch dial (often with a silvered disc between or an engraved 'sunburst').

Star wheel *See* Wheels.

Stop piece The lever that engages a hoop wheel or the pin on a stop wheel to end the operation of striking.

Stopwork This limits the working of a spring so that it cannot be wound past a set position, nor wind down past a set position. It is often found set in the end of the going barrel, when it consists of a specialised gear with only part of its circumference toothed and a corresponding single tooth gear that indexes the first by one space at every revolution (Geneva stopwork).

Striking and chiming Striking is the term applied to the use of bells or gongs to mark time, and utilises no more than two notes or one chord. Chiming is a succession of notes to form a tune; the tune may indicate the quarters by its content as in Westminster or Whittington chimes. A clock that plays a melody at the hour is called a musical clock. Popular songs of the day often feature in musical clocks, and these may be used to date at least part of the mechanism.

Supplementary arc The swing of a pendulum or balance wheel beyond what is necessary to operate the escapement. In the case of a recoil escapement it is when recoil takes place. Supplementary arc in a dead-beat escapement should result in no movement of the escape wheel.

Suspension The flexible support of a pendulum, verge balance wheel or foliot (not all verge balances and foliots have a flexible support at their upper end).

Swaging Squeezing metal to obtain a riveted joint.

Train A series of meshing gears.

Turnings The decorative rings around winding holes and the seatings of Brocot pallets.

Verge The arbor or staff that crosses the face of a crown wheel in a verge escapement. In America, it also refers to the bent-strip pallet and wire that is fitted to the front-mounted escape wheel of a recoil escapement.

Verge flags The pallets of a verge which accept impulse from the teeth of a crown wheel.

Warning The small rotation that the train of a striking or chiming clock makes before being released to complete the strike or chime. There is no warning in a repeater because this would prevent the clock indicating time after warning had taken place.

Wheels The large gears of a clock train. Except for very early clocks these are of wood or, more frequently, brass. Wooden wheels feature in early American thirty-hour clocks (the first mass-produced clock), some provincial British clocks and many early Black Forest clocks.

Centre wheel: The wheel on the arbor that carries the hands.

Contrate wheel: A wheel that carries gear

teeth at right angles to its face.
Crown wheel: An escape wheel that carries escapement teeth at right angles to its face.
Date wheel: A wheel turning once a day and operating a date ring.
Escape wheel: This rotates through the pallets of an escapement and is the last wheel in the going train.
Great wheel: The wheel that drives the train.
Hammer wheel: The wheel that lifts the hammer for striking.
Hoop wheel: The wheel that carries a ring with a space in it to engage the stopping piece.
Intermediate wheel: A wheel introduced between the great wheel and the centre wheel to extend the number of turns made in driving the centre wheel. Used in going-barrel clocks and long-duration clocks.
Minute wheel: The wheel engaged by the cannon pinion and associated with driving the hour hand.
Star wheel: This usually has twelve points or limbs. It carries a snail and enables the movement to change the time being struck precisely on the hour or the quarter, or whatever is required. Wheels of star form are used to lift the hammer on modern clocks.
Stop wheel: The wheel that carries a pin to engage the stop piece.
Third wheel: This is next to the centre wheel, and usually drives the escape wheel.
Warning wheel: This carries a pin which is used to hold back the train after warning, and which is released to complete the strike or chime.

Suppliers and Further Reading

GENERAL MATERIALS AND TOOLS

UK

Southern Watch and Clock Supplies Ltd, High St, Orpington, Kent BR6 0JH

Meadows and Passmore, Farningham Road, Crowborough, East Sussex TN6 2JP

Mahoney Assocs, 58 Stapleton Road, Bristol 5

H. S. Walsh and Sons Ltd, 243 Beckenham Road, Beckenham, Kent BR3 4TS

Nathan Shestopal Ltd, Unit 2, Sapcote Trading Centre, 374 High Rd, Willesden NW10

A. G. Thomas (Bradford) Ltd, Tompion House, Heaton Road, Bradford BD8 8RB

R. E. Rose, 731 Sidcup Road, Eltham, London SE9

A. Shoot and Sons Ltd, Renata House, 116–118 St John Street, London EC1

J.M.W. (Clocks), 12 Norton Green Close, Sheffield, Yorks S8 8BP

Chronos, 95 Victoria Street, St Albans, Herts

P. P. Thornton Successors Ltd, Old Bakehouse, Upper Tysoe, Warwickshire CV35 0TR

Geoffrey Booth, Tower House, Tower Hill, Bere Regis, Wareham, Dorset BH20 7JA

Richards of Burton, Woodhouse Clock Works, Swadlincote Road, Woodville, Burton-upon-Trent DE11 8DA

Collins, 99 Venable Avenue, Colne, Lancs BB8 7DH

Jones and Chambault, Gronfa, Station Road, Clynderwen, Dyfed SA66 7NF

Laurie Penman, Castle Workshop, High Street, Totnes, Devon TQ9 5PB

USA

The Cas-Ker Company, P.O. Box 2347, 128 East 6th Street, Cincinnatti, OH 45201

Esslinger & Company, 1165 Medallion Drive, St. Paul, MN 55120

The Gould Company, 13750 Neutron Road, Dallas, TX 75234

Herr & Kline Inc., 1914 Granby Street, Norfolk, VA 23517

S. LaRose, Inc., 234 Commerce Place, Greensboro, NC 27420

Marshall-Swartchild Company, 2040 North Milwaukee Avenue., Chicago, IL 60647

Mason and Sullivan Company, 586 Higgins Crowell Road, West Yarmouth, Cape Cod, MA 02673

The Nest Company, 915 Olive Street, St. Louis, MO 63101

E & J Swigart Company, 34 West Sixth Street, Cincinnatti, OH 45202

Tani Engineering, 6226 Waterloo, Box 338, Atwater, OH 44201

Additional supplier listings may be found in the *Watch and Clockmakers' Buyer's Guide* published by the American Watchmakers Institute, 3700 Harrison Avenue, Cincinnati, Ohio 45211.

DIAL RESTORATION

Rob Gillies, 38 Maltravers, Arundel, West Sussex

John Pearson, Church Cottage, Birstwith, Harrogate, Yorkshire

John Sawyer, 78 Bicester Road, Kidlington, Oxford

John E. Peters, 3 Arthur Road, Rainham, Kent ME8 9BT

Steve Collis, 58 Appleton Road, South Benfleet, Southend

Alan J. Thom, 55 Athens Street, Stockport SK1 4EG

Clive and Lesley Cobb, Newhouse Farm, Bratton Fleming, Barnstaple EX31 43T

Laurie Penman, Castle Workshop, High St, Totnes, Devon TQ9 5PB

FURTHER READING

UK

Britten, F. J. *Watch and Clockmaker's Handbook* (Eyre Methuen, 1978)

De Carle, D. *Practical Clock Repairing* (NAG, 1968)

Gazeley, W. J. *Watch and Clock Making and Repairing* (Newnes Butterworth, 1971)

Penman, L. *Clock Design and Construction* (Argus, 1984)

Rawlings, A. L. *Science of Clocks and Watches* (EP, 1974)

Wild, J. M. *Clock Wheel and Pinion Cutting* (Argus, 1983)

USA

Harris, H. R. *Nineteenth Century American Clocks* (Emerson Books, 1981)
Questions and Answers of and for the Clockmaking Profession (AWI Press)

Monk, Sean C. *Essence of Clock Repair* (AWI Press)

Palmer, B. *Treasury of American Clocks* (Macmillan Publishing, 1967)

Rudolph, J. S. *Build Your Own Working Clock* (Harper and Row, 1983)

Tyler, E. J. *American Clocks for the Collector* (E. P. Dutton, 1981)

Whiten, A. J. *Repairing Old Clocks and Watches* (Van Nostrand Rheinhold, 1981)

Other books available from American Watchmaker's Institute, P.O. Box 11011, Cincinnati, Ohio 45211. Contact them for a current catalog.

Index

Numbers in *italic* refer to figure numbers